EXCEL Statistics

EXCEL Statistics

A Quick Guide

Neil J. Salkind

University of Kansas

Los Angeles | London | New Delhi
Singapore | Washington DC

Los Angeles | London | New Delhi
Singapore | Washington DC

FOR INFORMATION:

SAGE Publications, Inc.
2455 Teller Road
Thousand Oaks, California 91320
E-mail: order@sagepub.com

SAGE Publications Ltd.
1 Oliver's Yard
55 City Road
London EC1Y 1SP
United Kingdom

SAGE Publications India Pvt. Ltd.
B 1/I 1 Mohan Cooperative Industrial Area
Mathura Road, New Delhi 110 044
India

SAGE Publications Asia-Pacific Pte. Ltd.
3 Church Street
#10-04 Samsung Hub
Singapore 049483

Acquisitions Editor: Vicki Knight
Assistant Editor: Kalie Koscielak
Production Editor: Libby Larson
Copy Editor: Paula L. Fleming
Typesetter: C&M Digitals (P) Ltd.
Proofreader: Wendy Jo Dymond
Indexer: Terri Corry
Cover Designer: Bryan Fishman
Marketing Manager: Nicole Elliott
Permissions Editor: Adele Hutchinson

Printed in the United States of America

Library of Congress Cataloging-in-Publication Data

Salkind, Neil J.

Excel statistics : a quick guide / Neil J. Salkind. — 2nd ed.

p. cm.
Includes bibliographical references and index.

ISBN 978-1-4522-5792-1 (pbk.)

1. Microsoft Excel (Computer file) 2. Social sciences—Statistical methods—Computer programs. 3. Electronic spreadsheets. I. Title.

HF5548.4.M523S263 2013
005.54—dc23 2012037167

This book is printed on acid-free paper.

12 13 14 15 16 10 9 8 7 6 5 4 3 2 1

Contents

How to Use This Book

Who *Excel Statistics: A Quick Guide* Is For . . .

Excel Statistics: A Quick Guide is about how to use the functions (which are predefined formulas) and the Analysis ToolPak available in Excel 2010. It was written for several general audiences.

For those enrolled in introductory statistics courses, *Excel Statistics: A Quick Guide* provides experience with the world's most popular spreadsheet program and how its features can be used to answer both simple and complex questions about data. *Excel Statistics: A Quick Guide* can also serve as an ancillary text to help students understand how an application such as Excel complements the introductory study of statistics.

For those who are more familiar with statistics, perhaps students in their second course or social or behavioral scientists or business researchers, *Excel Statistics: A Quick Guide* can be used to concisely show how functions, and the Analysis ToolPak tools can be used in applied situations. These users might consider *Excel Statistics: A Quick Guide* to be a reference book from which they can pick and choose the functions or tools they need to learn as they have the need.

Excel Statistics: A Quick Guide does not teach the reader how to use Excel as a spreadsheet application. It is assumed that the user of the book has some familiarity with basic computer operations (such as clicking and dragging) and has some knowledge of how to use Excel (such as how to enter and edit data and save files). Should you desire an introductory book that combines both the basics of statistics plus the use of Excel (beyond what is offered here), you might want to look at Salkind's *Statistics for People Who (Think They) Hate Statistics: Excel 2010 Edition*, also published by SAGE.

How to Use Excel Statistics: A Quick Guide

Excel Statistics: A Quick Guide is designed in a very special way.

Each of the 35 functions in Part 1 covers a two-page spread. On the left-hand side of the page is text, and on the right-hand side of the page are the figures (two per QuickGuide) that illustrate the ideas introduced in the text. This design and this format allow the user to see what the use of the Excel function looks like as it is being applied. It's easy to go from the text to the illustrations and back again if necessary.

Each of the 15 Analysis ToolPak tools in Part 2 uses a two-page spread as well and two figures—except that sometimes one of the figures appears on the left-hand page along with the text. This is because much of the resulting output takes up a considerable amount of physical page space and simply needs more room to be fully appreciated and understood.

With the above in mind, please note the following:

- If you are new to formulas or functions or the Analysis ToolPak, spend some time reading the introductions to Part 1 and Part 2 on pages 1 and 79. They will get you started quickly with using these Excel features.
- If you are familiar with functions and the Analysis ToolPak, then start looking through the functions and tools to get an idea of what is available and how the material is organized.
- Each of the functions and tools is accompanied by an Excel file that is used as an example. These files are available in two places: the SAGE website at www.sagepub.com/salkindexcelstats2e and the author's personal website at www.onlinefilefolder.com. At the latter site, look for the Excel QG folder; the username is *ancillaries*, and the password is *files.* Please contact me at njs@ku.edu should you have any difficulty downloading these files, and I will send them to you immediately.
- *Excel Statistics: A Quick Guide* is as much a reference book as anything else, and you should feel free to experiment with functions and tools that serve the same general purpose (such as the CORREL function and the Correlation ToolPak tool) to determine which work best for you in which situations and use all of Excel's very powerful features. For example, much of the output you see in *Excel Statistics: A Quick Guide* has been reformatted to better fit the page and look more attractive. This can be accomplished easily through the use of built-in Excel features such as table formatting. Excel is a very flexible tool, and you can use the output you generate through functions or through the use of tools in many different ways and in many different settings.

About the Windows and Macintosh Versions

The latest Windows and Macintosh versions of Excel are identical in many ways. The screens' appearance may appear just a bit different, but that is as much a function of the operating system as it is of the Excel application. There are clear (but minor) differences between the two in how you navigate around an Excel window and how you perform certain types of operations.

For example, in Windows you can copy a cell's contents by using the Ctrl+C key combination, whereas with the Mac you use the Command (Apple)+C key combination. Anyone familiar with either operating system will understand these differences, and you can feel comfortable using either version of the program. Once you know how to use the basics of Excel, the learning curve for the other version is not very steep.

However, the two versions do differ in some significant ways.

As far as functions go, the Mac version uses a simple and efficient Formula Builder to construct formulas and use functions, while the Windows version uses a similar (but a bit less friendly) Insert Function tool on the Formulas tab. At the same time, a significant disadvantage of the Mac version is that there is no Analysis ToolPak. If you want to do a more advanced analysis, like the ones we illustrate in Part II of *Excel Statistics: A Quick Guide*, then download StatPlus:mac LE (free from AnalystSoft), available as this book goes to press at www.analystsoft.com/en/products/statplusmacle/. Then use StatPlus:mac LE with the Mac version of Excel 2011.

Also, every few years, Microsoft, the developer of Excel, releases a new version of Excel. Usually, these new versions offer new and additional features, and just as often, many existing features stay the same. For the most part, the functions that you see in these pages work the same way as the functions in Microsoft 97–2003 and even earlier versions of Excel 5.0/95. So, although the images in these pages and the described steps are accurate and were developed using Excel 2010 for Windows (and 2011 for the Mac), you should have no concerns that these QuickGuides will not work with your version of Excel and with future versions such as Excel 2013. Also, with the introduction of both the Windows and Mac versions of Excel, Microsoft added some variants of formulas. However, there are corresponding functions and Analysis ToolPak tools that will work with any version, and those are the ones that we cover.

About the Second Edition

This edition differs from the first edition in two significant ways (other than what was previously discussed).

First, each Excel QuickGuide ends with two hands-on questions that require the reader to actually use data (available for download as noted earlier). These simple and direct exercises demonstrate how Excel functions and tools are used.

Second, Part I now contains five new functions that deal with working data—further enabling the reader to grasp how to organize and analyze information.

Acknowledgments

There are always many people to thank when a book comes to fruition. Let me first thank Vicki Knight, publisher at SAGE, for her editorship, guidance, patience, and willingness to talk through lots of aspects of the project. Lauren Habib, digital content editor, and Kalie Koscielak, assistant editor, are to be most gratefully thanked as well. Others to thank are Libby Larson for making this book look as good as it does (even if the author had a few crazy comments) and, as always, Paula Fleming, superb copy and development editor, who helped me answer difficult questions about how to illustrate data and what's best to include. And, thanks to all the people at SAGE who seem to care as much about their authors as what their authors write.

Corrections? Ideas? Concerns? Anything—let me know. Thanks.

Neil J. Salkind

njs@ku.edu

About the Author

Neil J. Salkind received his PhD in human development from the University of Maryland and taught at the University of Kansas for 35 years, where he is now professor emeritus. He has written more than 125 professional papers and presentations, and he is the author of several college-level textbooks, including *Statistics for People Who (Think They) Hate Statistics* (SAGE), *Theories of Human Development* (SAGE), *Exploring Research* (Prentice Hall), and the *Encyclopedia of Measurement and Statistics* (SAGE). Other SAGE texts by Salkind can be found at www.sagepub.com. He was also the editor of *Child Development Abstracts and Bibliography*. Salkind lives in Lawrence, Kansas, where he likes to read, swim with the River City Sharks, do letterpress printing, bake brownies (see the Excel version of *Statistics for People* . . . for the recipe), and poke around old Volvos.

PART I

USING EXCEL FUNCTIONS

A function is a formula that is predefined to accomplish a certain task. In this introduction, we will show the general steps for using any Excel function.

Although there are many different types of functions (such as AVERAGE and STDEV.S) in many different categories (such as financial, logical, and engineering), most of what we will be dealing with in *Excel Statistics: A Quick Guide* focuses on working with numerical data and performing elementary and advanced operations. In other words, we'll be learning to use those functions that fall in the group of functions known as *statistical*. Toward the end of Part I, we will introduce some database functions because they can be of particular interest, and importance, in analyzing data. They also fall under the general statistical functions category.

A Simple Example

In Figure I.1, you see a column of 10 numbers, with the average appearing in Cell B11. Take a look at the formula bar at the top of the figure, and you will see the syntax for this function:

=AVERAGE(B1:B10)

Figure I.1 A Simple Function That Computes the Average of a Set of Values

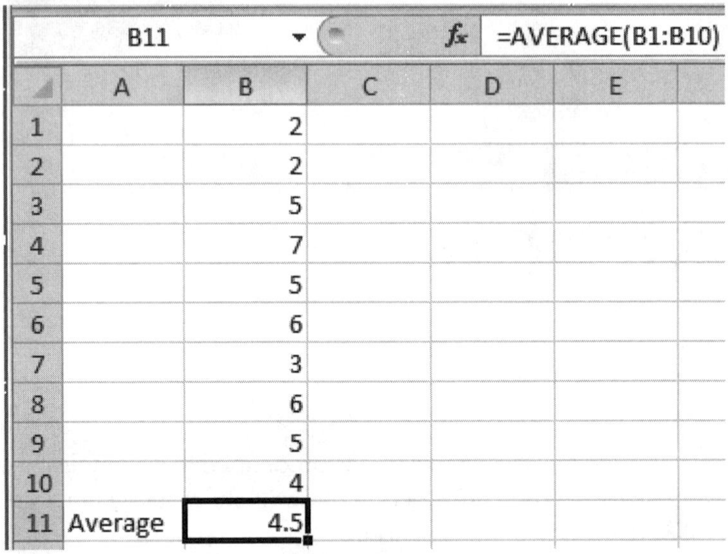

B11 =AVERAGE(B1:B10)

	A	B	C	D	E
1		2			
2		2			
3		5			
4		7			
5		5			
6		6			
7		3			
8		6			
9		5			
10		4			
11	Average	4.5			

Some general things to remember about using functions:

1. Functions always begin with an equal (=) sign as the first character entered in a cell.
2. A function can be placed in any cell. It will return the value of that function in that cell. It often makes sense to place the function near the data you are describing.
3. As you can see in Figure I.1, the syntax of the function, which is =AVERAGE(B1:B10), is not what is returned to the cell. Instead, you see the value computed by the function (in this case, 4.5) in the cell.
4. A function can be entered manually by typing the syntax and the range (or ranges) of cells that are to be applied), or it can be entered automatically. We'll review both methods in this introduction, but *Excel Statistics: A Quick Guide* will focus exclusively on the automatic method because it is easier and faster and it can always be edited manually if necessary.

The Anatomy of a Function

Let's take a quick look at what makes up a function, and then we will move on to how to enter one.

Here's the function that you saw in Figure I.1 that computes the average for a range of scores:

=AVERAGE(B1:B10)

And here is what each element represents . . .

=	Tells Excel that this is a function. You want Excel to calculate a value in the cell, not put the syntax in the cell
AVERAGE	Is the name of the function
(B1:B10)	Is the range of cells that hold the input to the function. In the case of the AVERAGE function, these cells hold the values you want to find the average of

Entering a Function Manually

To enter an Excel function manually (by typing an equal sign and the name of the function and the cells you want to be acted upon), you have to know two things:

1. The name of the function
2. The syntax or structure of the function. In the example that we have been using, AVERAGE is the name of the function and the syntax is =AVERAGE(range of cells).

Some functions are simple, and others are quite complex. You can find out the names of all available Excel functions, what they do, and their associated syntax through Excel Help.

Entering a Function Automatically

Entering a function automatically is by far the fastest and easiest method, and we'll use this model throughout *Excel Statistics: A Quick Guide*. Just as with entering a function manually, you need to know the function's name. In our ongoing example, we use the AVERAGE function and compute the actual value.

1. Click on the cell where you want the function to be placed. In our example, the cell is B11. If you are following along, be sure you click in a cell that is blank.

2. Click the Formulas tab and then click the Insert Function button (*fx*), and you will see the Insert Function dialog box as shown in Figure I.2.

Figure I.2 The Insert Function Dialog Box

Insert Function

Search for a function:

Type a brief description of what you want to do and then click Go

Go

Or select a category: Statistical

Select a function:

AVEDEV
AVERAGE
AVERAGEA
AVERAGEIF
AVERAGEIFS
BETA.DIST
BETA.INV

AVEDEV(number1,number2,...)

Returns the average of the absolute deviations of data points from their mean. Arguments can be numbers or names, arrays, or references that contain numbers.

Help on this function

OK Cancel

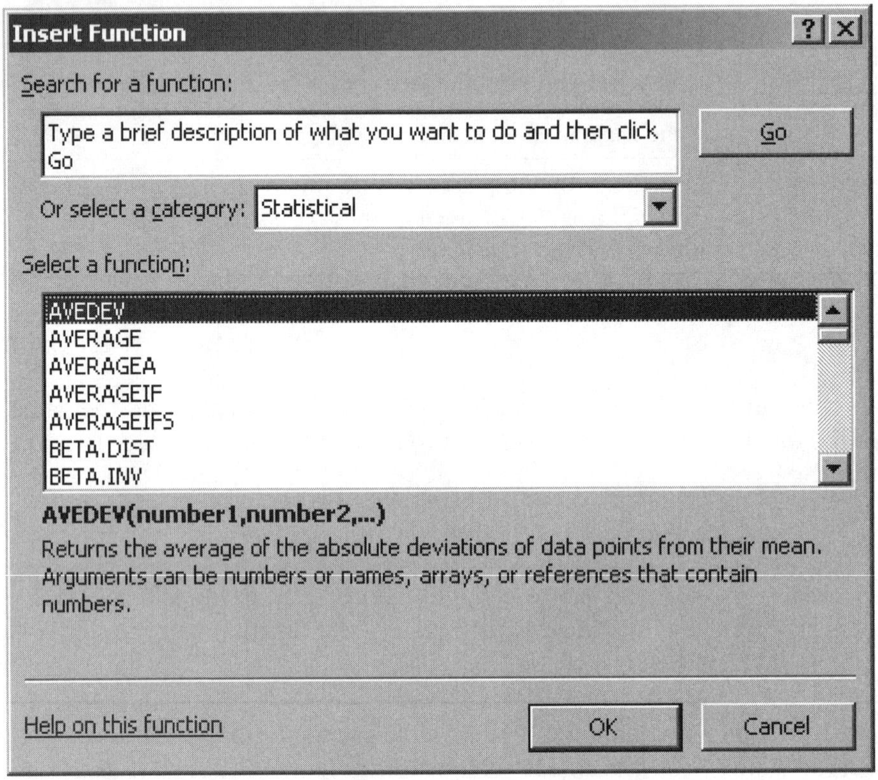

3. Locate the function you want to use by using the Search for a function box and the Go button, by selecting a category and then a function, or by selecting a function from the list that shows the most recently used functions. In our example, AVERAGE is the function of interest.

4. Double-click on the AVERAGE function in the list and you will see the all-important Function Arguments Box, shown in Figure I.3 with the cell addresses already completed. Note that when you enter a function in a cell below data, Excel assumes that you want the function to operate on the data in that column, so the appropriate cells

automatically appear as the range in the dialog box. If you have Excel enter a function and there is no data above that cell, then you will have to enter the range of data in the dialog box yourself.

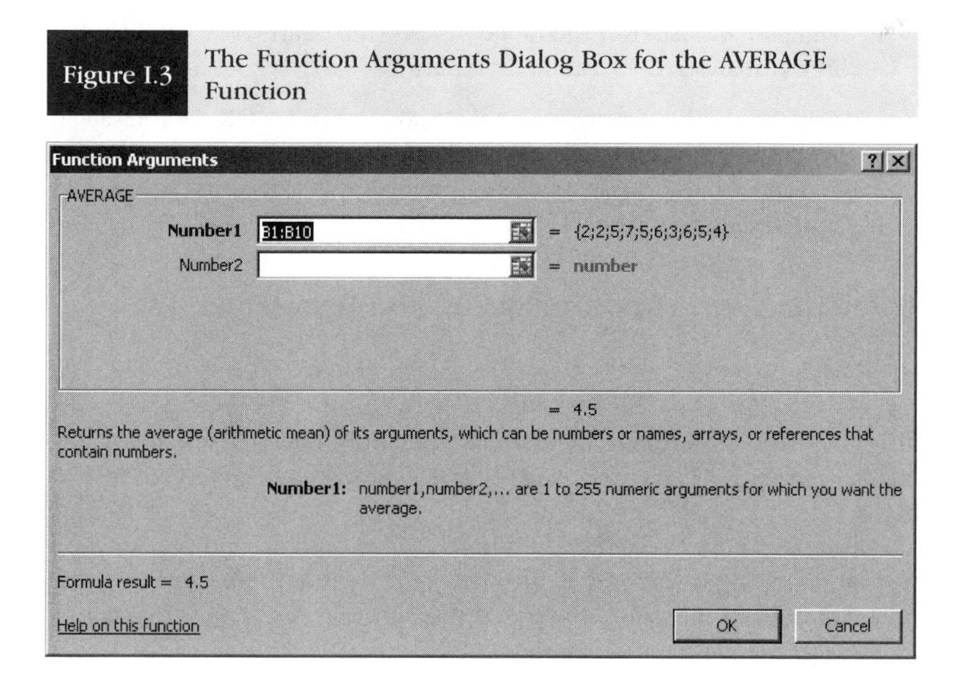

Figure I.3 The Function Arguments Dialog Box for the AVERAGE Function

Here's what's in the dialog box you see in Figure I.3. (Remember, because there is already a function in the box, you see cell references. If you clicked on a blank cell, no cells would have been entered yet.)

- There's the name of the function, AVERAGE.
- Then there are text boxes where you enter the range of cells (the argument) on which you want the function to perform its duty (B1:B10). Notice that the actual numbers (2, 2, 5, 7, 5, 6, 3, 6, 5, 4) you want to average are listed to the right of the text box.
- Right below the argument boxes is the value the function will return to the cell in which it is located (in this case, 4.5).
- Halfway down the dialog box is a description of what the function does (Returns the average . . .).
- Next is the syntax (or directions) of how to put the function together.
- The formula result (4.5) follows.
- Finally, there is a place to get help if you need it (*Help on this function*).

5. If Excel recognizes the data to which you want to apply the function, click OK and you are done. Otherwise, click on the RefEdit button and the Function Arguments box shrinks. (See the next section for an explanation of how the RefEdit button works.)

6. Drag the mouse over the range of cells (the data) you want included in the analysis. Click the RefEdit button.

7. Click the RefEdit button once again and then click OK. As you saw in Figure I.1, the results are shown in the Function Arguments box and returned to the worksheet. Also note that the syntax for the function appears in the formula bar.

About the RefEdit Button and Collapsing and Expanding

Microsoft (the developer of Excel) thought it best to refer to the collapse and expand button as the *RefEdit button*. When the Function Arguments dialog box is collapsed, clicking on the RefEdit button will expand the dialog box. And when it is expanded, clicking on the RefEdit button will collapse it. It's as simple as that.

In sum, you use the RefEdit button to collapse and expand the Function Arguments dialog box, and you are then able to drag the mouse over the cells that contain the information that the Function Arguments dialog box requires to complete the function.

Another Way to Enter Functions Automatically (Using a Few More Clicks Than the Above Method)

Another very useful way to have Excel help you insert a function is to use the Formulas tab and the More Functions option as follows:

1. Click the Formulas tab → More Functions → Statistical

2. Select the function you want to insert.

3. You will now see whatever Function Arguments dialog box is appropriate for the function you selected, and you can fill in the necessary information.

More About Function Names

As you use the newest version of Excel, you will see that the names of the possible functions appear in the cell as you type the function name. For example, when you type =STD . . . , you will see a drop box for the =STDEV.P function, for the STDEV.S function, and for several others (such as =STDEVA). In other words, Excel allows you to compute several types of standard deviations. We focus on those that we believe are most important and most often used. So, in Chapter 2, you will see QuickGuides for STDEV.S (#5) and STDEV.P (#6) but not for other standard deviation functions.

Second, certain functions are available in Excel but do not appear in the list of functions that appears in the Insert Function dialog box. For example, RANK.AVG and RANK.EQ both appear in the Insert Function dialog box, but RANK does not. But RANK still appears when you begin to type =RAN . . . in any cell. That's because the Excel developers only want people who have used earlier editions of Excel (where RANK was the only choice) to use that function in the latest version of Excel without distinguishing between RANK.AVG and RANK.EQ.

EXCEL QUICKGUIDE 1

The AVERAGE Function

What the AVERAGE Function Does

The AVERAGE function takes a set of values and computes the arithmetic mean, which is the sum of the values divided by the frequency of those values. It is the most often used measure of central tendency.

The Data Set

The data set used in this example is titled AVERAGE, and the question is, "What is the average speed of response and accuracy?"

Variable	*Description*
Response Time	Speed of response across 10 items
Accuracy	Number of items correct

Using the Average Function

1. Click on the cell where you want the AVERAGE function to be placed. (In the data set, the cell is B13.)
2. Click the Formulas tab → Insert Function button (*fx*) and locate and double-click the AVERAGE function. You will see the Function Arguments dialog box, as shown in Figure 1.1. Note that because the function is being inserted immediately below the data, Excel automatically recognizes what data should be included.
3. Click OK. The AVERAGE function returns its value in Cell B13, as you see in Figure 1.2. Copy the function to Cell C13. The average response time is 6.92, and the average accuracy is 7.60. Note that you can see the syntax for the function in the formula bar at the top of the worksheet.

Related Functions: MEDIAN, MODE.SNGL, GEOMEAN

Figure 1.1 The AVERAGE Function Arguments Dialog Box

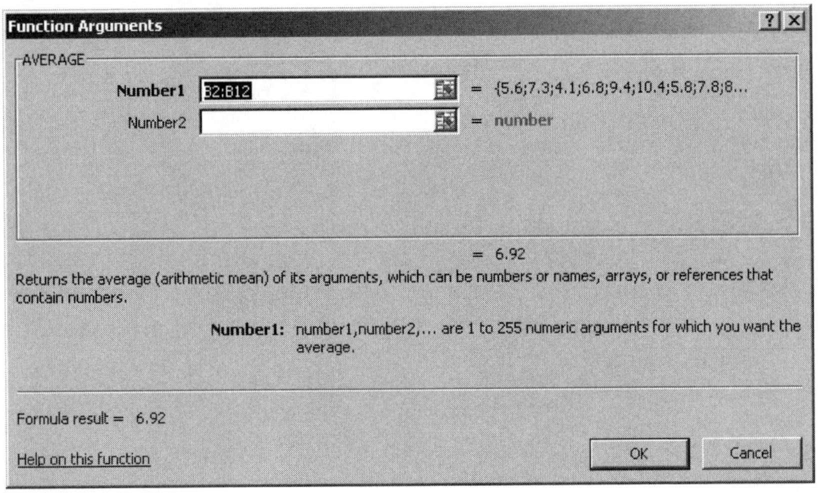

Function Arguments

AVERAGE

Number1 B2:B12 = {5.6;7.3;4.1;6.8;9.4;10.4;5.8;7.8;8...

Number2 = number

= 6.92

Returns the average (arithmetic mean) of its arguments, which can be numbers or names, arrays, or references that contain numbers.

Number1: number1,number2,... are 1 to 255 numeric arguments for which you want the average.

Formula result = 6.92

Help on this function

OK Cancel

Figure 1.2 The AVERAGE Function Returning the Mean Value

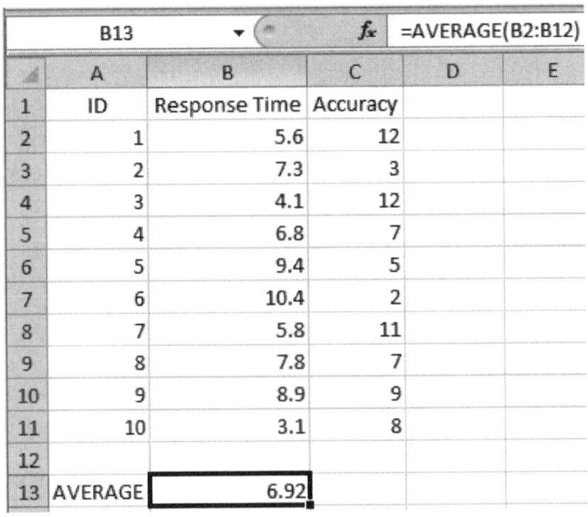

B13 *fx* =AVERAGE(B2:B12)

	A	B	C	D	E
1	ID	Response Time	Accuracy		
2	1	5.6	12		
3	2	7.3	3		
4	3	4.1	12		
5	4	6.8	7		
6	5	9.4	5		
7	6	10.4	2		
8	7	5.8	11		
9	8	7.8	7		
10	9	8.9	9		
11	10	3.1	8		
12					
13	AVERAGE	6.92			

Check Your Understanding

To check your understanding of the AVERAGE function, do the following two problems and check your answers in Appendix A.

QS 1a. Compute the average number of Sunday afternoon museum visitors across a 12-month period. Use Data Set 1a.

QS 1b. Compute the average age at which 20 undergraduates received their 4-year degree. Use Data Set 1b.

EXCEL QUICKGUIDE 2

The MEDIAN Function

What the MEDIAN Function Does

The MEDIAN function computes the data point at which 50% of the values fall above it and 50% of the values fall below it. It is most often used as a measure of central tendency when there are extreme scores.

The Data Set

The data set used in this example is titled MEDIAN, and the question is, "What is the median annual income for a group of 11 homeowners?"

Variable	*Description*
Income	Annual income in dollars

Using the MEDIAN Function

1. Click on the cell where you want the MEDIAN function to be placed. (In the data set, the cell is B14.)

2. Click the Formulas tab → Insert Function button (*fx*) and locate and double-click on the MEDIAN function. You will see the Function Arguments dialog box, as shown in Figure 2.1. Note that because the function is being inserted immediately below the data, Excel automatically recognizes what data should be included.

3. Click OK. The MEDIAN function returns its value in Cell B14, as you see in Figure 2.2. The median income value is $56,525. Note that you can see the syntax for the function in the formula bar at the top of the worksheet.

Related Functions: AVERAGE, MODE.SNGL, GEOMEAN

Figure 2.1 The MEDIAN Function Arguments Dialog Box

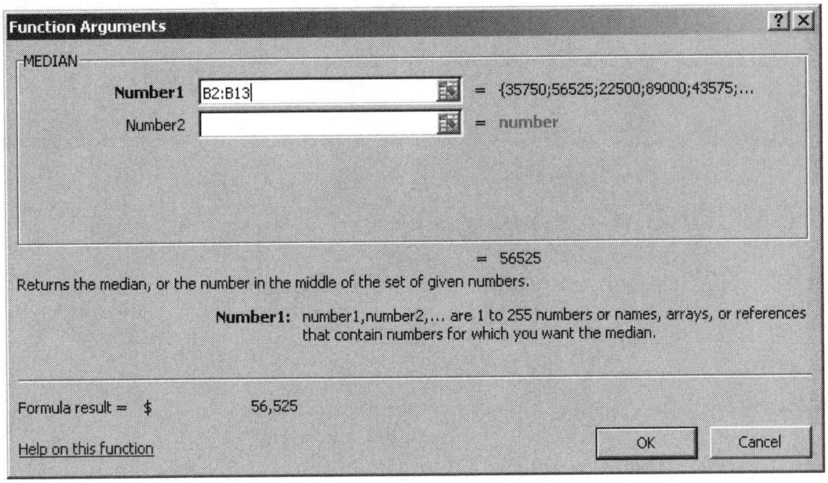

Function Arguments

MEDIAN

Number1 B2:B13 = {35750;56525;22500;89000;43575;...

Number2 = number

= 56525

Returns the median, or the number in the middle of the set of given numbers.

Number1: number1,number2,... are 1 to 255 numbers or names, arrays, or references that contain numbers for which you want the median.

Formula result = $ 56,525

Help on this function

OK Cancel

Figure 2.2 The MEDIAN Function Returning the Median Value

B14 =MEDIAN(B2:B13)

	A	B	C	D	E
1	ID	Income			
2	1	$ 35,750			
3	2	$ 56,525			
4	3	$ 22,500			
5	4	$ 89,000			
6	5	$ 43,575			
7	6	$ 21,000			
8	7	$ 59,000			
9	8	$ 71,250			
10	9	$354,000			
11	10	$ 54,250			
12	11	$ 65,500			
13					
14	MEDIAN	$ 56,525			

Check Your Understanding

To check your understanding of the MEDIAN function, do the following two problems and check your answers in Appendix A.

QS 2a. Compute the median price of a home across 10 different communities measured in dollars. Use Data Set 2a.

QS 2b. Compute the median height of a group of 20 NCAA basketball players measured in inches. Use Data Set 2b.

EXCEL QUICKGUIDE 3

The MODE.SNGL Function

What the MODE.SNGL Function Does

The MODE.SNGL function computes the most frequently occurring value in a set of values.

The Data Set

The data set used in this example is titled MODE, and the question is, "What is the mode, or favorite, flavor of ice cream?"

Variable	*Description*
Preference	1 = vanilla ice cream, 2 = strawberry ice cream, 3 = chocolate ice cream

Using the MODE.SNGL Function

1. Click on the cell where you want the MODE function to be placed. (In the data set, the cell is B23.)

2. Click the Formulas tab → Insert Function button (*fx*) and locate and double-click on the MODE.SNGL function. You will see the Function Arguments dialog box, as shown in Figure 3.1. Note that because the function is being inserted immediately below the data, Excel automatically recognizes what data should be included.

3. Click OK. The MODE.SNGL function returns its value in Cell B23, as you see in Figure 3.2. The MODE.SNGL is 3; in other words, the most popular flavor of ice cream is chocolate. Note that you can see the syntax for the function in the formula bar at the top of the worksheet.

Related Functions: AVERAGE, MEDIAN, GEOMEAN

Figure 3.1 The MODE.SNGL Function Arguments Dialog Box

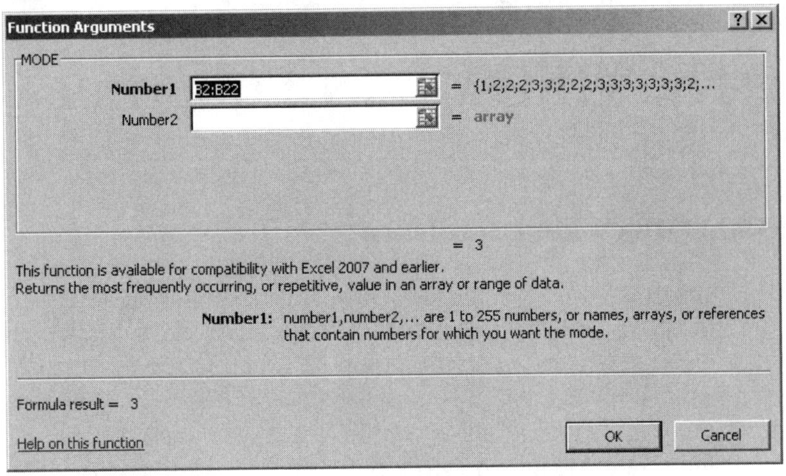

Function Arguments

MODE

Number1 B2:B22 = {1;2;2;2;3;3;2;2;2;3;3;3;3;3;3;3;2;...

Number2 = array

= 3

This function is available for compatibility with Excel 2007 and earlier.
Returns the most frequently occurring, or repetitive, value in an array or range of data.

Number1: number1,number2,... are 1 to 255 numbers, or names, arrays, or references that contain numbers for which you want the mode.

Formula result = 3

Help on this function

OK Cancel

Figure 3.2 The MODE.SNGL Function Returning the Mode Value

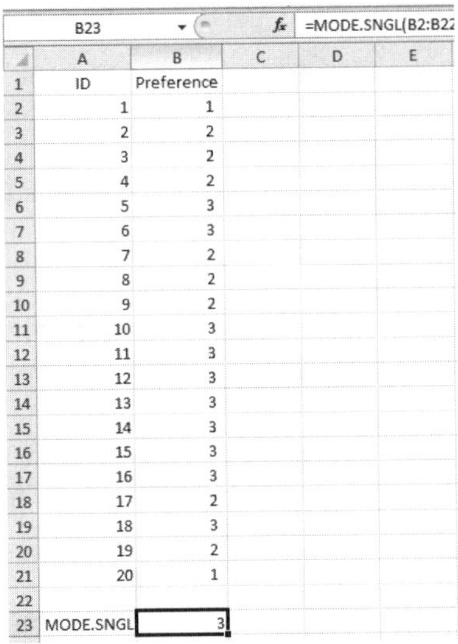

B23 fx =MODE.SNGL(B2:B22

	A	B	C	D	E
1	ID	Preference			
2	1	1			
3	2	2			
4	3	2			
5	4	2			
6	5	3			
7	6	3			
8	7	2			
9	8	2			
10	9	2			
11	10	3			
12	11	3			
13	12	3			
14	13	3			
15	14	3			
16	15	3			
17	16	3			
18	17	2			
19	18	3			
20	19	2			
21	20	1			
22					
23	MODE.SNGL	3			

Check Your Understanding

To check your understanding of the MODE.SNGL function, do the following two problems and check your answers in Appendix A.

QS 3a. Compute the mode for toppings for a group of students who order ice cream sundaes. Use Data Set 3a, where 1 = hot fudge, 2 = butterscotch, and 3 = no preference.

QS 3b. Compute the mode for age range of 25 respondents on a survey. Use Data Set 3b, where 1 = below 21 years of age, 2 = between 21 and 34, and 3 = over 34 years of age.

EXCEL QUICKGUIDE 4

The GEOMEAN Function

What the GEOMEAN Function Does

The GEOMEAN function computes the geometric mean. It multiplies a set of numbers and takes the *n*th root of the product, where *n* is the number of numbers in the set. The geometric mean is generally used to compute an appropriate weighted average when a factor, such as reading ability, is measured on two different scales, say reading speed in words per minute and recall of what was read on a scale of 1 to 5. As another example, the United Nations Human Development Index uses the geometric mean.

The Data Set

The data set used in this example is titled GEOMEAN, and the question is, "What is the geometric mean for percent gains in achievement over a 5-year period?"

Variable	*Description*
Ach	Achievement gain for years 1 through 5

Using the GEOMEAN Function

1. Click on the cell where you want the GEOMEAN function to be placed. (In the data set, the cell is B8.)
2. Click the Formulas tab → Insert Function button (*fx*) and locate and double-click on the GEOMEAN function. You will see the Function Arguments dialog box, as shown in Figure 4.1. Note that because the function is being inserted immediately below the data, Excel automatically recognizes what data should be included.
3. Click OK. The GEOMEAN function returns its value as .083 or 8.3% in Cell B8, as you see in Figure 4.2. Note that you can see the syntax for the function in the formula bar at the top of the worksheet.

Related Functions: AVERAGE, MEDIAN, MODE.SNGL

Figure 4.1 The GEOMEAN Function Arguments Dialog Box

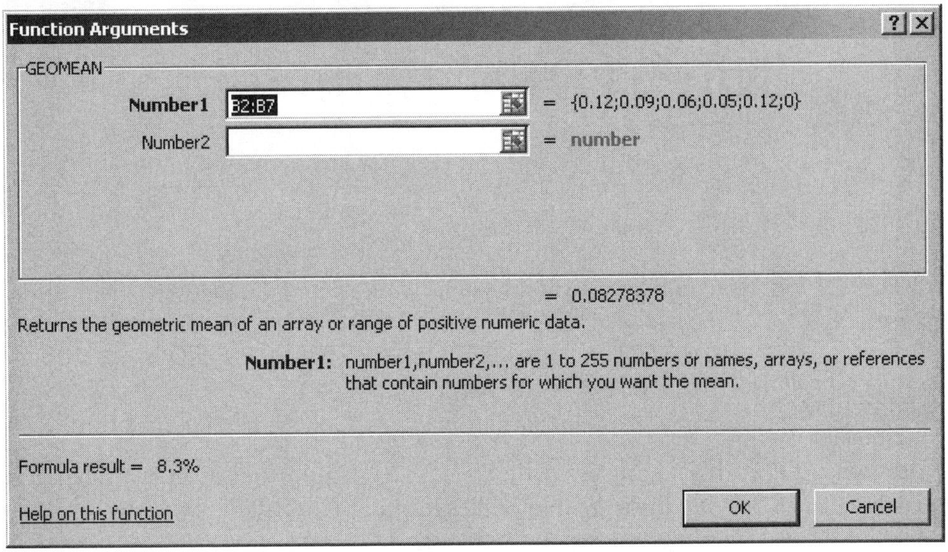

Function Arguments

GEOMEAN

Number1 B2:B7 = {0.12;0.09;0.06;0.05;0.12;0}

Number2 = number

= 0.08278378

Returns the geometric mean of an array or range of positive numeric data.

Number1: number1,number2,... are 1 to 255 numbers or names, arrays, or references that contain numbers for which you want the mean.

Formula result = 8.3%

Help on this function

OK Cancel

Figure 4.2 The GEOMEAN Function Returning the Geometric Mean Value

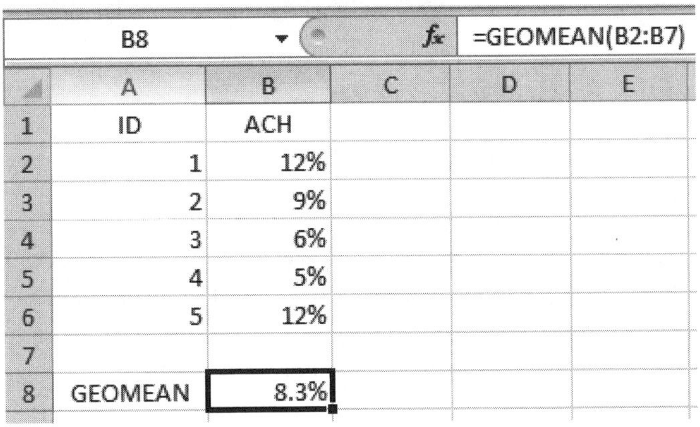

B8 fx =GEOMEAN(B2:B7)

	A	B	C	D	E
1	ID	ACH			
2	1	12%			
3	2	9%			
4	3	6%			
5	4	5%			
6	5	12%			
7					
8	GEOMEAN	8.3%			

Check Your Understanding

To check your understanding of the GEOMEAN function, do the following two problems and check your answers in Appendix A.

QS 4a. Compute the geometric mean for average achievement score for a group of 10 high schools. Use Data Set 4a.

QS 4b. Compute the geometric mean for average amount spent on groceries per week in dollars over a year. Use Data Set 4b.

EXCEL QUICKGUIDE 5

The STDEV.S Function

What the STDEV.S Function Does

The STDEV.S function takes a sample set of values and computes the standard deviation.

The Data Set

The data set used in this example is titled STDEV.S, and the question is, "What is the standard deviation for age for a sample of 20 sixth graders?"

Variable	*Description*
Age	Age in months for a group of sixth graders

Using the STDEV.S Function

1. Click on the cell where you want the STDEV.S function to be placed. (In the data set, the cell is B23.)

2. Click the Formulas tab → Insert Function button (*fx*) and locate and double-click on the STDEV.S function. You will see the Function Arguments dialog box, as shown in Figure 5.1. Note that because the function is being inserted immediately below the data, Excel automatically recognizes what data should be included.

3. Click OK. The STDEV.S function returns its value in Cell B23, as you see in Figure 5.2. The standard deviation for the sample's age is 4.32. Note that you can see the syntax for the function in the formula bar at the top of the worksheet.

Related Functions: STDEV.P, VAR.S, VAR.P

Figure 5.1 The STDEV.S Function Arguments Dialog Box

Function Arguments

STDEV.S

Number1 B2:B22 = {144;145;151;146;146;153;144;154...

Number2 = number

= 4.319783133

Estimates standard deviation based on a sample (ignores logical values and text in the sample).

Number1: number1,number2,... are 1 to 255 numbers corresponding to a sample of a population and can be numbers or references that contain numbers.

Formula result = 4.32

Help on this function

OK Cancel

Figure 5.2 The STDEV.S Function Returning the Standard Deviation for the Sample

B23 =STDEV.S(B2:B22)

	A	B	C	D	E
1	ID	Age in Months			
2	1	144			
3	2	145			
4	3	151			
5	4	146			
6	5	146			
7	6	153			
8	7	144			
9	8	154			
10	9	148			
11	10	154			
12	11	141			
13	12	145			
14	13	149			
15	14	139			
16	15	140			
17	16	143			
18	17	147			
19	18	146			
20	19	143			
21	20	145			
22					
23	STDEV.S	4.32			

Check Your Understanding

To check your understanding of the STDEV.S function, do the following two problems and check your answers in Appendix A.

QS 5a. Compute the standard deviation for percent correct on a sample of 10 spelling test scores. Use Data Set 5a.

QS 5b. Compute the standard deviation for the number of adults who received a flu shot across five different Midwest counties. Use Data Set 5b.

EXCEL QUICKGUIDE 6

The STDEV.P Function

What the STDEV.P Function Does

The STDEVP function computes the standard deviation for scores from an entire population.

The Data Set

The data set used in this example is titled STDEV.P, and the question is, "What is the standard deviation for the number of friendships lasting more than 1 year for a population of 90 tenth graders?"

Variable	*Description*
Number of Friends	• Number of friendships lasting more than 1 year for a group of tenth graders.

Using the STDEV.P Function

1. Click on the cell where you want the STDEV.P function to be placed. (In the data set, the cell is B18.)
2. Click the Formulas tab → Insert Function button (*fx*) and locate and double-click on the STDEV.P function. You will see the Function Arguments dialog box, as shown in Figure 6.1. You must enter the range of B2:G16 (instead of B2:B17) in the Number1 text box. Note that because the data is not in one column, Excel does not automatically enter the proper range.
3. Click OK. The STDEV.P function returns its value in Cell B18, as you see in Figure 6.2. The standard deviation for number of friendships for the population is 1.98. Note that you can see the syntax for the function in the formula bar at the top of the worksheet.

Related Functions: STDEV.S, VAR.S, VAR.P

Figure 6.1 The STDEV.P Function Arguments Dialog Box

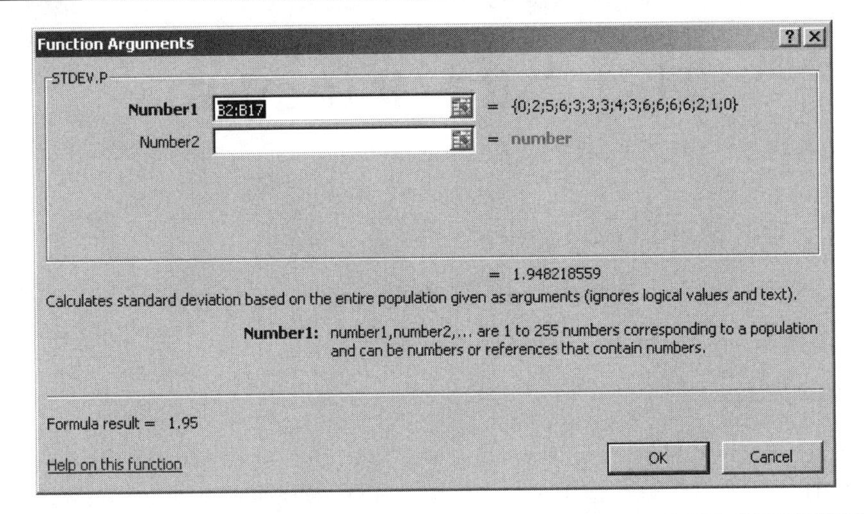

Figure 6.2 The STDEV.P Function Returning the Standard Deviation for the Population

B18 | =STDEV.P(B2:G16)

	A	B	C	D	E	F	G
				Number of Friends			
1							
2		3	1	2	3	6	5
3		6	2	2	2	1	3
4		0	3	2	0	0	1
5		4	5	0	1	3	1
6		0	5	4	6	6	2
7		6	2	5	4	4	6
8		4	1	5	5	6	3
9		3	2	6	6	3	6
10		0	3	1	1	6	2
11		4	1	4	2	4	4
12		5	2	4	5	6	5
13		0	5	0	0	2	3
14		3	5	1	5	5	6
15		1	4	0	2	6	3
16		4	1	3	1	5	1
17							
18	STDEV.P	1.98					

Check Your Understanding

To check your understanding of the STDEV.P function, do the following two problems and check your answers in Appendix A.

QS 6a. Compute the standard deviation for the population of community center monthly users. Use Data Set 6a.

QS 6b. Compute the standard deviation for the population of shoppers at Hobbs toy store during one week in spring. Use Data Set 6b.

EXCEL QUICKGUIDE 7

The VAR.S Function

What the VAR.S Function Does

The VAR.S function computes the variance for a set of sample scores.

The Data Set

The data set used in this example is titled VAR, and the question is, "What is the variance for the scores on The Extroversion Scale (TES) for a sample of 20 high school seniors?"

Variable	*Description*
TES	Score on The Extroversion Scale ranging from 1 to 100

Using the VAR.S Function

1. Click on the cell where you want the VAR.S function to be placed. (In the data set, the cell is B23.)

2. Click the Formulas tab → Insert Function button (*fx*) and locate and double-click on the VAR.S function. You will see the Function Arguments dialog box, as shown in Figure 7.1. Note that because the function is being inserted immediately below the data, Excel automatically recognizes what data should be included.

3. Click OK. The VAR.S function returns its value in Cell B23, as you see in Figure 7.2. The variance of this TES scores for this sample is 246.37. Note that you can see the syntax for the function in the formula bar at the top of the worksheet.

Related Functions: STDEV.S, STDEV.P, VAR.P

Figure 7.1 The VAR.S Function Arguments Dialog Box

Function Arguments

VAR.S

Number1 B2:B22 = {78;98;85;37;67;74;56;69;82;58;75...

Number2 = number

= 246.3657895

Estimates variance based on a sample (ignores logical values and text in the sample).

Number1: number1,number2,... are 1 to 255 numeric arguments corresponding to a sample of a population.

Formula result = 246.37

Help on this function

OK Cancel

Figure 7.2 The VAR.S Function Returning the Variance for the Sample

B23 f_x =VAR.S(B2:B22)

	A	B	C	D	E
1	ID	TES			
2	1	78			
3	2	98			
4	3	85			
5	4	37			
6	5	67			
7	6	74			
8	7	56			
9	8	69			
10	9	82			
11	10	58			
12	11	75			
13	12	69			
14	13	78			
15	14	37			
16	15	69			
17	16	77			
18	17	89			
19	18	65			
20	19	76			
21	20	90			
22					
23	VAR.S	246.37			

Check Your Understanding

To check your understanding of the VAR.S function, do the following two problems and check your answers in Appendix A.

QS 7a. Compute the variance for a sample of 20 scores, ranging from 1 to 100, on a mathematics test. Use Data Set 7a.

QS 7b. Compute the variance for a sample of 10 scores, ranging in value from 1 to 10, that measure aggressive behavior. Use Data Set 7b.

EXCEL QUICKGUIDE 8

The VAR.P Function

What the VAR.P Function Does

The VAR.P function computes the variance for an entire population.

The Data Set

The data set used in this example is titled VAR.P, and the question is, "What is the variance for the scores on The Extroversion Scale (TES) for the entire population of 100 high school seniors?"

Variable	*Description*
TES	Score on The Extroversion Scale ranging from 1 to 100

Using the VAR.P Function

1. Click on the cell where you want the MEDIAN function to be placed. (In the data set, the cell is B23.)

2. Click the Formulas tab → Insert Function button (*fx*) and locate and double-click on the VAR.P function. You will see the Function Arguments dialog box, as shown in Figure 8.1. Enter the range of A2:E21 (instead of B2:B22) in the Number1 text box. Note that because the data is not in one column, Excel does not automatically enter the proper range.

3. Click OK. The VAR.P function returns its value in Cell B23, as you see in Figure 8.2. The variance of TES scores for the population is 684.12. Note that you can see the syntax for the function in the formula bar at the top of the worksheet.

Related Functions: STDEV.S, STDEV.P, VAR.S

Figure 8.1 The VAR.P Function Arguments Dialog Box

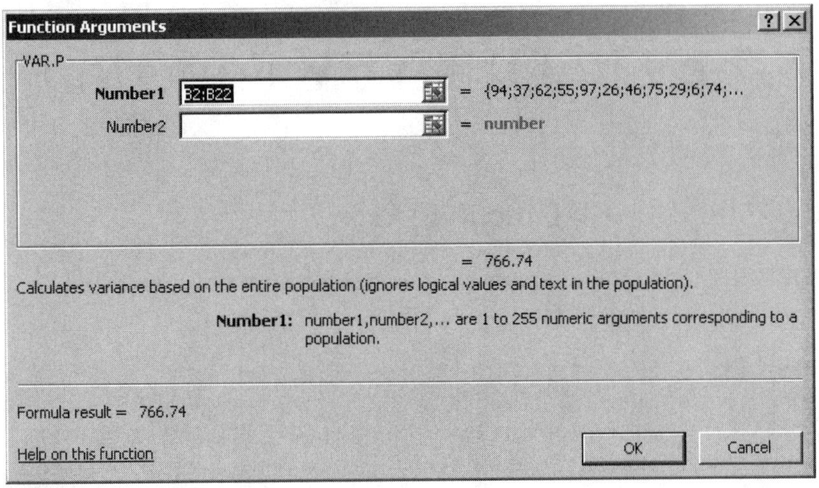

Function Arguments

VAR.P

Number1 B2:B22 = {94;37;62;55;97;26;46;75;29;6;74;...

Number2 = number

= 766.74

Calculates variance based on the entire population (ignores logical values and text in the population).

Number1: number1,number2,... are 1 to 255 numeric arguments corresponding to a population.

Formula result = 766.74

Help on this function

OK Cancel

Figure 8.2 The VAR.P Function Returning the Variance for the Population

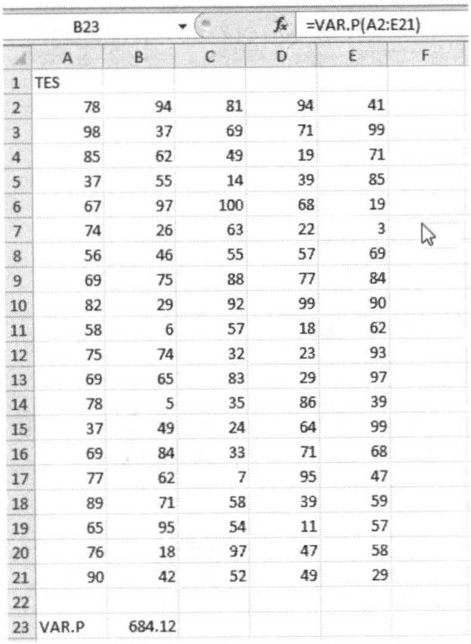

B23 =VAR.P(A2:E21)

	A	B	C	D	E	F
1	TES					
2	78	94	81	94	41	
3	98	37	69	71	99	
4	85	62	49	19	71	
5	37	55	14	39	85	
6	67	97	100	68	19	
7	74	26	63	22	3	
8	56	46	55	57	69	
9	69	75	88	77	84	
10	82	29	92	99	90	
11	58	6	57	18	62	
12	75	74	32	23	93	
13	69	65	83	29	97	
14	78	5	35	86	39	
15	37	49	24	64	99	
16	69	84	33	71	68	
17	77	62	7	95	47	
18	89	71	58	39	59	
19	65	95	54	11	57	
20	76	18	97	47	58	
21	90	42	52	49	29	
22						
23	VAR.P	684.12				

Check Your Understanding

To check your understanding of the VAR.P function, do the following two problems and check your answers in Appendix A.

QS 8a. Compute the variance for the population of annual income for all breadwinners in a county of 10 cities. Use Data Set 8a.

QS 8b. Compute the variance for the population of Body Mass Index (BMI) scores for all 100 students in an elementary school. Use Data Set 8b.

EXCEL QUICKGUIDE 9

The FREQUENCY Function

What the FREQUENCY Function Does

The FREQUENCY function generates frequencies for a set of values.

The Data Set

The data set used in this example is titled FREQUENCY, and the question is, "What is the frequency of level of preference for TOOTS, a new type of cereal, by 60 consumers?"

Variable	*Description*
Preference	Preference on a scale from 1 through 5

Using the FREQUENCY Function

1. Create the bins (1, 2, 3, 4, and 5) into which you want the frequencies counted (as shown in Cells I2 through I6 in the data set).
2. Highlight all the cells (J2 through J6) where you want the frequencies to appear (see Figure 9.2).
3. Click the Formulas tab → Insert Function button (*fx*) and locate and double-click on the FREQUENCY function. You will see the Function Arguments dialog box, as shown in Figure 9.1.
4. Click the RefEdit button in the Data_array entry box and drag the mouse over the range of cells you want included in the analysis (B2 through G11). Click the RefEdit button.
5. Click the RefEdit button in the Bins_array entry box and drag the mouse over the range of cells that defines the bins (I2 through I6). Click the RefEdit button.
6. Press the Ctrl–Shift–Enter keys in combination (not in sequence). This is done because Excel treats Cells J2 through J6 as an array, not as a single value. In Figure 9.2, you can see the result of the function with frequencies listed beside the values in the bin. Note in the formula bar that the function is bounded by brackets {}, indicating the values are part of an array.

Related Functions: NORM.DIST, PERCENTILE.INC, PERCENTRANK.INC, QUARTILE.INC, RANK.AVG, STANDARDIZE

Figure 9.1 The FREQUENCY Function Arguments Dialog Box

Function Arguments

FREQUENCY

Data_array = reference

Bins_array = reference

=

Calculates how often values occur within a range of values and then returns a vertical array of numbers having one more element than Bins_array.

Data_array is an array of or reference to a set of values for which you want to count frequencies (blanks and text are ignored).

Formula result =

Help on this function

OK Cancel

Figure 9.2 The FREQUENCY Function Returning the Frequencies of Values

J2 *fx* {=FREQUENCY(B2:G11,I2:I6)}

	A	B	C	D	E	F	G	H	I	J
1		Preference							Bins	FREQUENCY
2		2	1	1	1	3	5		1	10
3		2	2	4	5	4	1		2	14
4		2	3	5	2	1	3		3	11
5		5	2	3	1	4	2		4	8
6		3	2	5	5	5	3		5	17
7		4	4	1	1	5	5			
8		1	4	2	3	2	3			
9		5	3	4	4	2	2			
10		5	2	5	5	5	1			
11		2	5	5	3	5	3			

Check Your Understanding

To check your understanding of the FREQUENCY function, do the following two problems and check your answers in Appendix A.

QS 9a. Generate a frequency histogram for the classes of patients according to the following condition codes: 1 = ambulatory; 2 = nonambulatory. Use Data Set 9a.

QS 9b. Generate a frequency histogram for the classes of consumers according to the following condition codes: 1 = loved the service; 2 = didn't like the service; 3 = couldn't care less. Use Data Set 9b.

EXCEL QUICKGUIDE 10

The NORM.DIST Function

What the NORM.DIST Function Does

The NORMDIST function computes the cumulative probability of a score.

The Data Set

The data set used in this example is titled NORM.DIST, and the question is, "What is the cumulative probability associated with a spelling test score of 9?"

Variable	*Description*
Spelling Score	Number of words spelled correctly out of 20

Using the NORM.DIST Function

To use the NORMDIST function, follow these steps:

1. Click on the cell where you want the NORM.DIST function to be placed. (In the data set, the cell is C2.)
2. Click the Formulas tab → Insert Function button (*fx*) and locate and double-click on the NORM.DIST function. You will see the Function Arguments dialog box, as shown in Figure 10.1.
4. For X, click the RefEdit button and enter the location of the value for which you want to compute the probability. (In the data set, it is B2.)
5. For the Mean and Standard_dev, click the RefEdit button and enter the appropriate cell addresses (B14 for the Mean and B15 for the Standard_dev). Note that you need to have these values already computed before you can use this function.
6. For Cumulative, type True.
7. Click OK, and the cumulative probability will be computed in Cell C2. In this example, the cumulative probability associated with a spelling score of 9 is 0.01 or 1%.
8. Edit the function in cell C2 to include absolute references for cells B14 and B15, as you see in Figure 10.2 ($B$14 and $B$15). Copy this result from Cell C3 through Cell C12, and the results of the probabilities associated with scores from 9 through 19 will appear, as shown in Figure 10.2.

Related Functions: FREQUENCY, PERCENTILE.INC, PERCENTRANK.INC, QUARTILE.INC, RANK.AVG, STANDARDIZE

Figure 10.1 The NORM.DIST Function Arguments Dialog Box

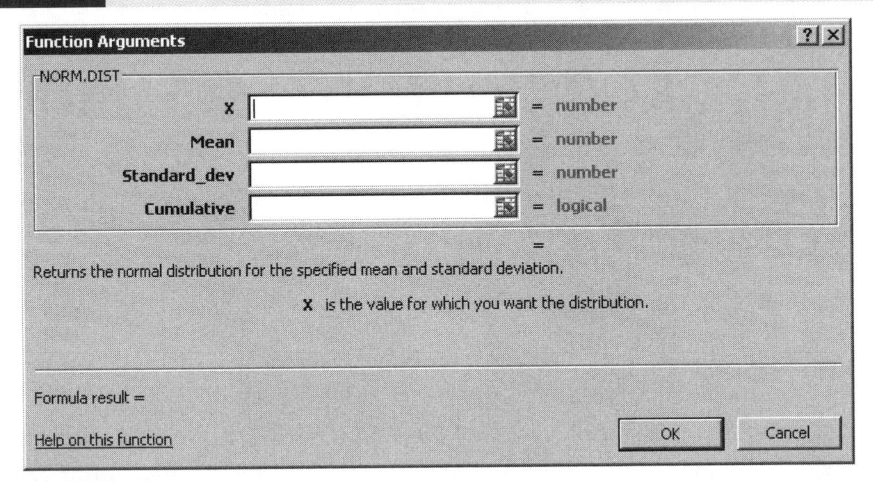

Figure 10.2 The NORM.DIST Function Returning the Cumulative Probability for a Set of Spelling Scores

C2 =NORM.DIST(B2,$B$14,$B$15,TRUE)

	A	B	C	D
1		Spelling Score	NORM.DIST (Cumulative Probability)	
2		9	1%	
3		10	2%	
4		11	7%	
5		12	16%	
6		13	31%	
7		14	50%	
8		15	69%	
9		16	84%	
10		17	93%	
11		18	98%	
12		19	99%	
13				
14	Class Mean	14		
15	Class sd	2		

Check Your Understanding

To check your understanding of the FREQUENCY function, do the following two problems and check your answers in Appendix A.

QS 10a. Compute the cumulative probabilities for sales by month. Use Data Set 10a.

QS 10b. Compute the cumulative probabilities for number of cars in a dealer's inventory by quarter. Use Data Set 10b.

EXCEL QUICKGUIDE 11

The PERCENTILE.INC Function

What the PERCENTILE.INC Function Does

The PERCENTILE.INC function computes the value for a defined percentile.

The Data Set

The data set used in this example is titled PERCENTILE.INC, and the question is, "What are the percentile values for strength, as measured by weight lifted, in a sample of twenty-five 55-year-olds?

Variable	*Description*
Strength	Amount of weight lifted

Using the PERCENTILE.INC Function

1. Click on the cell where you want the PERCENTILE.INC function to be placed. (In the data set, the cell is D2.)
2. Click the Formulas tab → Insert Function button (*fx*) and locate and double-click on the PERCENTILE.INC function. You will see the Function Arguments dialog box, as shown in Figure 11.1.
3. Click the RefEdit button in the Array entry box and drag the mouse over the range of cells (B2 through B26) you want included in the analysis. Click the RefEdit button.
5. Click the RefEdit button in the K entry box and click on the cell (C2) for which you want to compute the percentile value.
6. Click the RefEdit button and press the Enter key or click OK. The PERCENTILE.INC function returns the value 171 in Cell D2. The percentile function was then copied from Cell D2 through Cell D11, as you see in Figure 11.2. Note that you can see the syntax for the function in the formula bar at the top of the worksheet.

Related Functions: FREQUENCY, NORM.DIST, PERCENTRANK.INC, QUARTILE.INC, RANK.AVG, STANDARDIZE

Figure 11.1 The PERCENTILE.INC Function Arguments Dialog Box

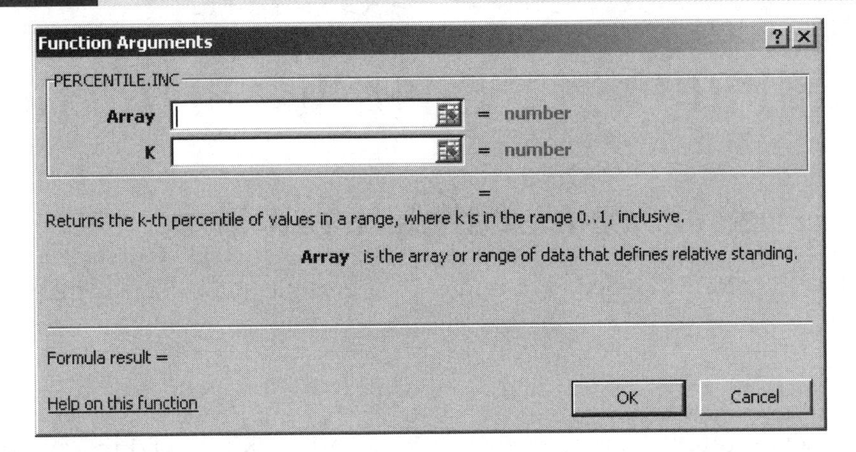

Figure 11.2 The PERCENTILE.INC Function Returning the Values Associated with Percentiles

D2 =PERCENTILE.INC($B$2:$B$26,C2)

	A	B	C	D	E	F	G
1	ID	Strength	Percentile	PERCENTILE.INC (Value)			
2	1	149	100%	171.0			
3	2	142	90%	157.6			
4	3	117	80%	149.6			
5	4	132	70%	147.0			
6	5	152	60%	145.4			
7	6	171	50%	134.0			
8	7	134	40%	128.2			
9	8	166	30%	122.2			
10	9	101	20%	119.4			
11	10	147	10%	109.4			
12	11	148					
13	12	116					
14	13	105					
15	14	147					
16	15	157					
17	16	120					
18	17	122					
19	18	120					
20	19	146					
21	20	158					
22	21	100					
23	22	123					
24	23	131					
25	24	124					
26	25	145					

Check Your Understanding

To check your understanding of the PERCENTILE.INC function, do the following two problems and check your answers in Appendix A.

QS 11a. In the set of data for use of library facilities by weekday for the entire month of June, what is the value of the 50th percentile? Use Data Set 11a.

QS 11b. In the set of data for number of meals prepared in facilities by month for the entire year, what is the value of the 80th percentile? Use Data Set 11b.

EXCEL QUICKGUIDE 12

The PERCENTRANK.INC Function

What the PERCENTRANK.INC Function Does

The PERCENTRANK.INC function computes the percentile rank of a particular value in a data set.

The Data Set

The data set used in this example is titled PERCENTRANK.INC, and the question is, "What is the percentile rank (or percentile) for the individual who lifts 149 pounds?"

Variable	*Description*
Strength	Amount of weight lifted

Using the PERCENTRANK.INC Function

1. Click on the cell where you want the PERCENTRANK.INC function to be placed. (In the data set, the cell is C2.)
2. Click the Formulas tab → Insert Function button (*fx*) and locate and double-click on the PERCENTRANK.INC function. You will see the Function Arguments dialog box as shown in Figure 12.1.
3. Click the RefEdit button in the Array entry box.
5. Drag the mouse over the range of cells (B2 through B26) you want included in the analysis.
6. Click the RefEdit button in the X entry box.
7. Click the X value (Cell B2) for which you want to compute the PERCENTRANK.INC or percentile rank.
8. Click the RefEdit button and press the Enter key or click OK. The PERCENTRANK.INC function returns the value 79% in Cell C2. The percentile values were then copied from Cell C2 through Cell C26 (after changing B2:B26 to $B$2:$B$26 so that absolute, and not relative references, are used), as you see in Figure 12.2. Note that you can see the syntax for the function in the formula bar at the top of the worksheet.

Related Functions: FREQUENCY, NORM.DIST, PERCENTILE.INC, QUARTILE.INC, RANK.AVG, STANDARDIZE

Figure 12.1 The PERCENTRANK.INC Function Arguments Dialog Box

Function Arguments

PERCENTRANK.INC

Array = number

X = number

Significance = number

=

Returns the rank of a value in a data set as a percentage of the data set as a percentage (0..1, inclusive) of the data set.

Array is the array or range of data with numeric values that defines relative standing.

Formula result =

Help on this function

OK Cancel

Figure 12.2 The PERCENTRANK.INC Function Returning the Percentile Ranks for the Set of Scores

C2 =PERCENTRANK.INC($B$2:$B$26,B2)

	A	B	C
1	ID	Strength	PERCENTRANK.INC (Value)
2	1	149	79%
3	2	142	54%
4	3	117	17%
5	4	132	46%
6	5	152	83%
7	6	171	100%
8	7	134	50%
9	8	166	96%
10	9	101	4%
11	10	147	67%
12	11	148	75%
13	12	116	13%
14	13	105	8%
15	14	147	67%
16	15	157	88%
17	16	120	21%
18	17	122	29%
19	18	120	21%
20	19	146	63%
21	20	158	92%
22	21	100	0%
23	22	123	33%
24	23	131	42%
25	24	124	38%
26	25	145	58%

Check Your Understanding

To check your understanding of the PERCENTRANK.INC function, do the following two problems and check your answers in Appendix A.

QS 12a. Compute the value of the percent rank for a test score of 89. Use Data Set 12a.

QS 12b. Compute the value of the percent rank for a house sale value of $156,500. Use Data Set 12b.

EXCEL QUICKGUIDE 13

The QUARTILE.INC Function

What the QUARTILE.INC Function Does

The QUARTILE.INC function computes the values that divide a set of data into quartiles or fourths.

The Data Set

The data set used in this example is titled QUARTILE.INC, and the question is, "What is the first quartile (or 25th percentile) in a set of health scores for a group of 25 nonsmokers?"

Variable	*Description*
Health Score	Risk of chronic illness score from 1 to 50

Using the QUARTILE.INC Function

1. Click on the cell where you want the QUARTILE.INC function to be placed. (In the data set, the cell is D2.)
2. Click the Formulas tab → Insert Function button (*fx*) and locate and double-click on the QUARTILE.INC function. You will see the Function Arguments dialog box, as shown in Figure 13.1.
3. Click the RefEdit button in the Array entry box.
4. Drag the mouse over the range of cells (B2 through B26) you want included in the analysis.
5. Click the RefEdit button in the Quart entry box.
6. Click the Quart value (Cell C2) to compute the 1st quartile.
7. Click the RefEdit button and press the Enter key or click OK. The QUARTILE.INC function returns the value 27 in Cell D2, as you can see in Figure 13.2. The 2nd, 3rd, and 4th quartiles were computed as well. Note that you can see the syntax for the function in the formula bar at the top of the worksheet.

Related Functions: FREQUENCY, NORM.DIST, PERCENTILE.INC, PERCENTRANK.INC, RANK.AVG, STANDARDIZE

Figure 13.1 The QUARTILE.INC Function Arguments Dialog Box

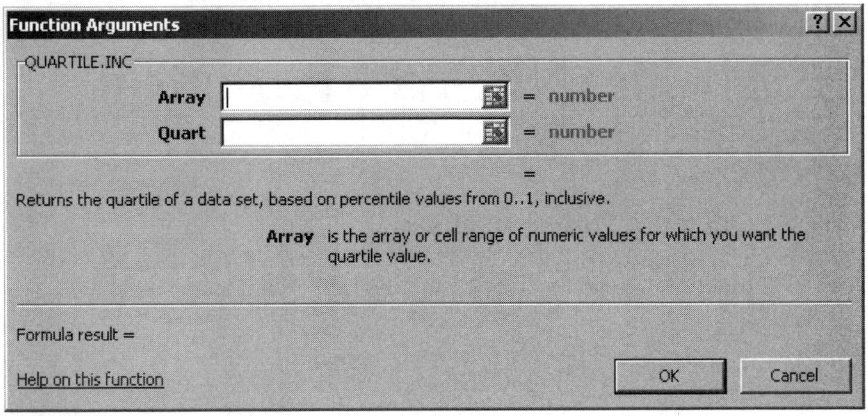

Function Arguments

QUARTILE.INC

Array = number

Quart = number

=

Returns the quartile of a data set, based on percentile values from 0..1, inclusive.

Array is the array or cell range of numeric values for which you want the quartile value.

Formula result =

Help on this function

OK Cancel

Figure 13.2 The QUARTILE.INC Function Returning the Quartiles for a Set of Scores

D2 =QUARTILE.INC($B$2:$B$26,C2)

	A	B	C	D	E	F
1	ID	Health Score	Quartile	QUARTILE.INC Value		
2	1	35	1	27		
3	2	46	2	33		
4	3	50	3	44		
5	4	32	4	50		
6	5	14				
7	6	4				
8	7	27				
9	8	37				
10	9	24				
11	10	31				
12	11	36				
13	12	43				
14	13	45				
15	14	48				
16	15	44				
17	16	31				
18	17	33				
19	18	27				
20	19	18				
21	20	2				
22	21	7				
23	22	44				
24	23	41				
25	24	28				
26	25	48				

Check Your Understanding

To check your understanding of the QUARTILE.INC function, do the following two problems and check your answers in Appendix A.

QS 13a. Compute the value of the 2nd quartile for a listing of 25 scores on a spelling test with 20 items. Use Data Set 13a

QS 13b. Compute the 3rd quartile for a set of health care costs at 20 area hospital emergency rooms. Use Data Set 13b.

EXCEL QUICKGUIDE 14

The RANK.AVG Function

What the RANK.AVG Function Does

The RANK.AVG function computes the percentage rank of a particular value in a data set.

The Data Set

The data set used in this example is titled RANK.AVG, and the question is, "What is the relative rank for a set of 20 grade point averages?"

Variable	*Description*
GPA	Grade point average

Using the RANK.AVG Function

1. Click on the cell where you want the RANK.AVG function to be placed. (In the data set, the cell is C2.)
2. Click the Formulas tab → Insert Function button (*fx*) and locate and double-click on the RANK.AVG function. You will see the Function Arguments dialog box, as shown in Figure 14.1.
3. Click on the cell where you want the RANK.AVG function to be placed. (In the data set, the cell is C2.)
4. Click the RefEdit button in the Number entry box and click the value for which you want to compute the rank (Cell B2).
5. Click the RefEdit button in the Ref entry box and drag the mouse over the array of values being ranked (B2 through B21).
6. Click the RefEdit button and press the Enter key or click OK. The RANK.AVG function returns the value 5 in Cell C2 (for a GPA of 3.4), as you see in Figure 14.2. The ranks for all the other data points are computed as well. Remember that when the ranks are copied, absolute references need to be used for the cell range B2 through B21 ($B$2:$B$21). Note that you can see the syntax for the function in the formula bar at the top of the worksheet.

Related Functions: FREQUENCY, NORM.DIST, PERCENTILE.INC, PERCENTRANK.INC, QUARTILE.INC, STANDARDIZE

Figure 14.1 The RANK.AVG Function Arguments Dialog Box

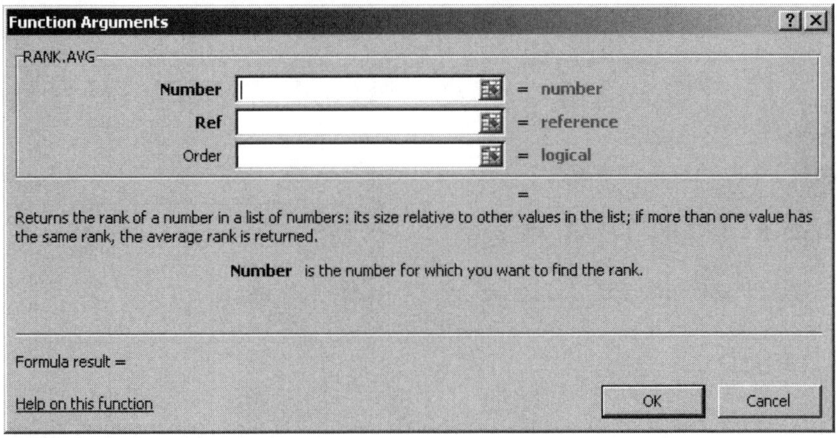

Function Arguments

RANK.AVG

Number = number

Ref = reference

Order = logical

=

Returns the rank of a number in a list of numbers: its size relative to other values in the list; if more than one value has the same rank, the average rank is returned.

Number is the number for which you want to find the rank.

Formula result =

Help on this function

OK Cancel

Figure 14.2 The RANK.AVG Function Returning the Ranks for a Set of Scores

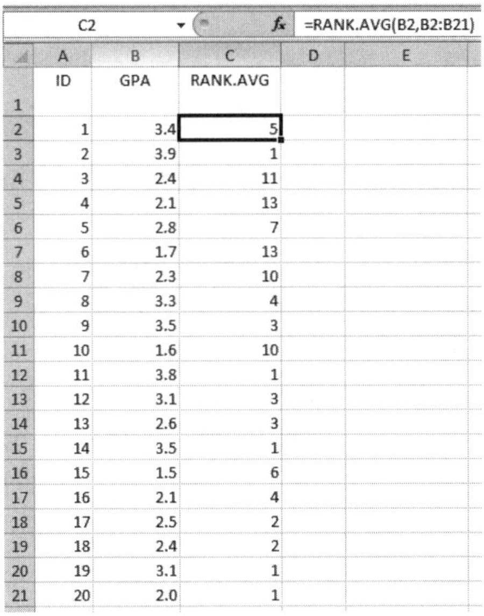

C2 =RANK.AVG(B2,B2:B21)

	A	B	C	D	E
1	ID	GPA	RANK.AVG		
2	1	3.4	5		
3	2	3.9	1		
4	3	2.4	11		
5	4	2.1	13		
6	5	2.8	7		
7	6	1.7	13		
8	7	2.3	10		
9	8	3.3	4		
10	9	3.5	3		
11	10	1.6	10		
12	11	3.8	1		
13	12	3.1	3		
14	13	2.6	3		
15	14	3.5	1		
16	15	1.5	6		
17	16	2.1	4		
18	17	2.5	2		
19	18	2.4	2		
20	19	3.1	1		
21	20	2.0	1		

Check Your Understanding

To check your understanding of the RANK.AVG function, do the following two problems and check your answers in Appendix A.

QS 14a. Compute the ranking for a score of 57 from a set of 50 response times, measured in seconds. Use Data Set 14a.

QS 14b. Compute the ranking for team #6 of 25 teams for total points scored in a season. Use Data Set 14b.

EXCEL QUICKGUIDE 15

The STANDARDIZE Function

What the STANDARDIZE Function Does

The STANDARDIZE function computes a normalized or standard score.

The Data Set

The data set used in this example is titled STANDARDIZE, and the question is, "What are the normalized scores for a set of raw test scores on the INT, a measure of introversion?"

Variable	*Description*
INT	A measure of introversion ranging from 1 to 25

Using the STANDARDIZE Function

1. Click on the cell where you want the STANDARDIZE function to be placed. (In the data set, the cell is C2.)
2. Click the Formulas tab → Insert Function button (*fx*) and locate and double-click on the STANDARDIZE function. You will see the Function Arguments dialog box, as shown in Figure 15.1.
3. Click the RefEdit button in the X entry box and click the value for which you want to compute STANDARDIZE (Cell B2). Click the RefEdit button.
4. Click the RefEdit button in the Mean entry box and click the mean of the values being standardized (Cell B23). Click the RefEdit button.
5. Click the RefEdit button in the Standard_dev entry box and click the standard deviation of the values being standardized (Cell B24). Click the RefEdit button.
6. Click the RefEdit button and press the Enter key or click OK. The STANDARDIZE function returns the value −0.58 in Cell C2, as you can see in Figure 15.2. The standardized scores for all the other data points were also computed (using relative references for the mean and standard deviation). The syntax for the function is shown in the formula bar at the top of the worksheet.

Related Functions: FREQUENCY, NORM.DIST, PERCENTILE.INC, PERCENTRANK.INC, QUARTILE.INC, RANK.AVG

Figure 15.1 The STANDARDIZE Function Arguments Dialog Box

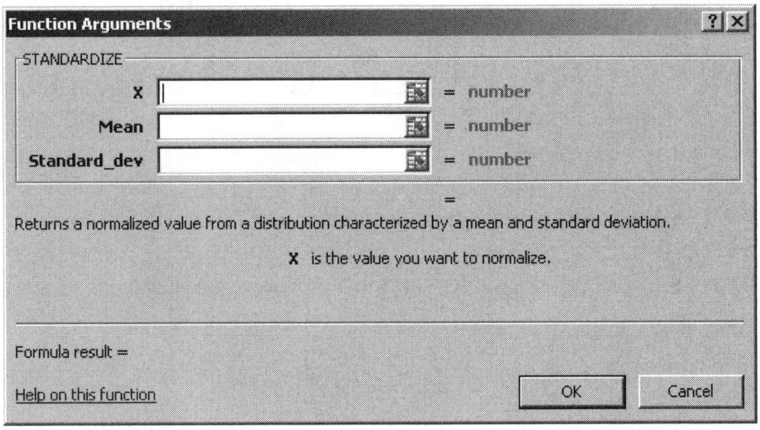

Function Arguments

STANDARDIZE

X = number

Mean = number

Standard_dev = number

=

Returns a normalized value from a distribution characterized by a mean and standard deviation.

X is the value you want to normalize.

Formula result =

Help on this function

OK Cancel

Figure 15.2 The STANDARDIZE Function Returning Standardized Values for a Set of Scores

C2 =STANDARDIZE(B2,$B$23,$B$24)

	A	B	C	D
1	ID	INT	STANDARDIZE	
2	1	12	-0.58	
3	2	15	-0.08	
4	3	14	-0.24	
5	4	21	0.93	
6	5	23	1.26	
7	6	25	1.59	
8	7	7	-1.41	
9	8	15	-0.08	
10	9	21	0.93	
11	10	8	-1.24	
12	11	22	1.09	
13	12	15	-0.08	
14	13	14	-0.24	
15	14	22	1.09	
16	15	19	0.59	
17	16	17	0.26	
18	17	9	-1.08	
19	18	10	-0.91	
20	19	3	-2.08	
21	20	17	0.26	
22				
23	AVERAGE	15.45		
24	STDEV.S	6		

Check Your Understanding

To check your understanding of the STANDARDIZE function, do the following two problems and check your answers in Appendix A.

QS 15a. Compute all the standardized scores for a set of 25 test scores from a history class for the fall semester. Note that you have to compute the mean and standard deviation. Use Data Set 15a.

QS 15b. Compute the standardized score for a raw score of 76 from set of 25 test scores from a history class for the spring semester. Use Data Set 15b.

EXCEL QUICKGUIDE 16

The COVARIANCE.S Function

What the COVARIANCE.S Function Does

The COVARIANCE.S function and estimates how much two variables change together.

The Data Set

The data set used in this example is titled COVARIANCE.S, and the question is, "What is the relationship between level of intervention (hours of training) and number of injuries in 20 college athletes?"

Variable	*Description*
Intervention	Hours of training
Injuries	Number of injuries

Using the COVARIANCE.S Function

1. Click on the cell where you want the COVARIANCE.S function to be placed. (In the data set, the cell is C23.)
2. Click the Formulas tab → Insert Function button (*fx*) and locate and double-click on the COVARIANCE.S function. You will see the Function Arguments dialog box, as shown in Figure 16.1.
3. Click the RefEdit button in the Array1 entry box and drag the mouse over the range of cells (B2 through B21) you want included in the analysis. Click the RefEdit button.
4. Repeat Step 3 for Array2 (Cells C2 through C21) and click the RefEdit button.
5. Press the Enter key or click OK. The COVARIANCE.S function returns its value as −1.9053 in Cell C23, as you see in Figure 16.2. You can see the syntax for the function in the formula bar at the top of the worksheet.

Related Functions: CORREL, PEARSON, INTERCEPT, SLOPE, TREND, FORECAST, RSQ

Figure 16.1 The COVARIANCE.S Function Arguments Dialog Box

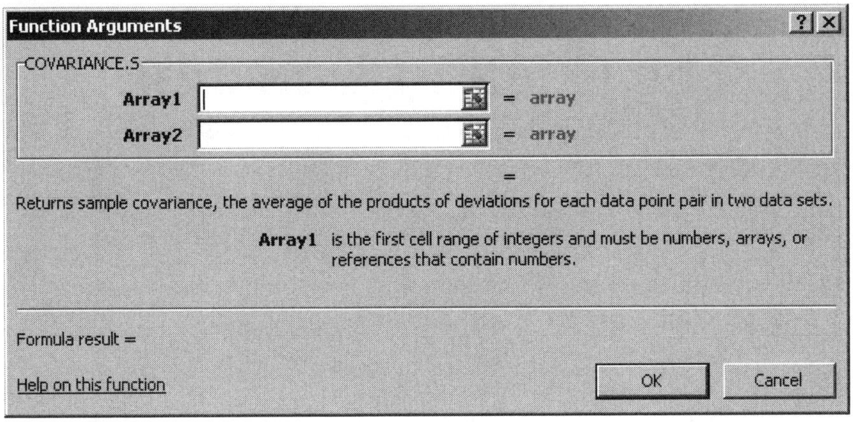

Function Arguments

COVARIANCE.S

Array1 = array

Array2 = array

=

Returns sample covariance, the average of the products of deviations for each data point pair in two data sets.

Array1 is the first cell range of integers and must be numbers, arrays, or references that contain numbers.

Formula result =

Help on this function

OK Cancel

Figure 16.2 The COVARIANCE.S Function Returning the Covariance

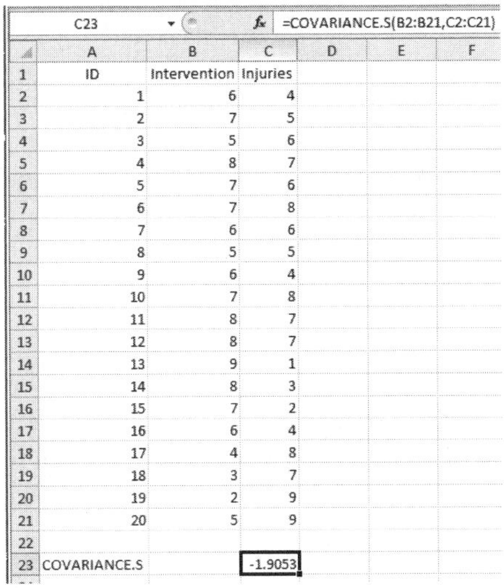

C23 =COVARIANCE.S(B2:B21,C2:C21)

	A	B	C	D	E	F
1	ID	Intervention	Injuries			
2	1	6	4			
3	2	7	5			
4	3	5	6			
5	4	8	7			
6	5	7	6			
7	6	7	8			
8	7	6	6			
9	8	5	5			
10	9	6	4			
11	10	7	8			
12	11	8	7			
13	12	8	7			
14	13	9	1			
15	14	8	3			
16	15	7	2			
17	16	6	4			
18	17	4	8			
19	18	3	7			
20	19	2	9			
21	20	5	9			
22						
23	COVARIANCE.S		-1.9053			

Check Your Understanding

To check your understanding of the COVARIANCE.S function, do the following two problems and check your answers in Appendix A.

QS 16a. Compute the covariance for time to first response and correct responses for 15 sixth graders. Use Data Set 16a.

QS 16b. Compute the median for a set of hours of weekly study time and GPA for 25 first-year college students. Use Data Set 16b.

EXCEL QUICKGUIDE 17

The CORREL Function

What the CORREL Function Does

The CORREL function computes the value of the Pearson product-moment correlation between two variables.

The Data Set

The data set used in this example is titled CORREL, and the question is, "What is the correlation between height and weight for 20 sixth graders?"

Variable	*Description*
Height	Height in inches
Weight	Weight in pounds

Using the CORREL Function

1. Click on the cell where you want the CORREL function to be placed. (In the data set, the cell is C23.)
2. Click the Formulas tab → Insert Function button (*fx*) and locate and double-click on the CORREL function. You will see the Function Arguments dialog box, as shown in Figure 17.1.
3. Click the RefEdit button in the Array1 entry box and drag the mouse over the range of cells (B2 through B21) you want included in the analysis. Click the RefEdit button.
4. Repeat Step 3 for Array2 (Cells C2 through C21) and then click the RefEdit button.
5. Click OK. The CORREL function returns its value as .77592 in Cell C23, as you see in Figure 17.2. You can see the syntax for the function in the formula bar at the top of the worksheet.

Related Functions: COVARIANCE.S, PEARSON, INTERCEPT, SLOPE, TREND, FORECAST, RSQ

Figure 17.1 The CORREL Function Arguments Dialog Box

Function Arguments

CORREL

Array1 = array

Array2 = array

=

Returns the correlation coefficient between two data sets.

Array1 is a cell range of values. The values should be numbers, names, arrays, or references that contain numbers.

Formula result =

Help on this function

OK Cancel

Figure 17.2 The CORREL Function Returning the Correlation

C23 =CORREL(B2:B21,C2:C21)

	A	B	C	D	E
1	ID	Height	Weight		
2	1	60	134		
3	2	63	143		
4	3	71	156		
5	4	58	121		
6	5	61	131		
7	6	59	117		
8	7	64	125		
9	8	67	126		
10	9	63	143		
11	10	52	98		
12	11	61	154		
13	12	58	125		
14	13	54	109		
15	14	61	117		
16	15	64	126		
17	16	63	154		
18	17	49	98		
19	18	59	143		
20	19	69	144		
21	20	71	156		
22					
23	CORREL		0.77592		

Check Your Understanding

To check your understanding of the CORREL function, do the following two problems and check your answers in Appendix A.

QS 17a. Compute the correlation coefficient for number of children retained in grade in 20 classrooms with parent's level of school involvement, with the latter measured on a scale from 1 to 10 with 10 being most involved. Use Data Set 17a.

QS 17b. Compute the correlation coefficient for number of water treatment plants and incidence of infectious diseases, with the latter measured by how many of 100 residents in 10 communities fall ill. Use Data Set 17b.

EXCEL QUICKGUIDE 18

The PEARSON Function

What the PEARSON Function Does

The PEARSON function computes the value of the Pearson product-moment correlation between two variables.

The Data Set

The data set used in this example is titled PEARSON, and the question is, "What is the correlation between income and level of education for 20 households?"

Variable	*Description*
Income	Annual income in dollars
Level of Education	Years of education

Using the PEARSON Function

1. Click on the cell where you want the PEARSON function to be placed. (In the data set, the cell is C23.)
2. Click the Formulas tab → Insert Function button (*fx*) and locate and double-click on the PEARSON function. You will see the Function Arguments dialog box, as shown in Figure 18.1.
3. Click the RefEdit button in the Array1 entry box and drag the mouse over the range of cells (B2 through B21) you want included in the analysis. Click the RefEdit button.
4. Repeat Step 3 for Array2 (Cells C2 through C21) and then click the RefEdit button.
5. Click OK. The PEARSON function returns its value as .53 in Cell C23, as you see in Figure 18.2. You can see the syntax for the function in the formula bar at the top of the worksheet.

Related Functions: COVARIANCE.S, CORREL, INTERCEPT, SLOPE, TREND, FORECAST, RSQ

Figure 18.1 The PEARSON Function Arguments Dialog Box

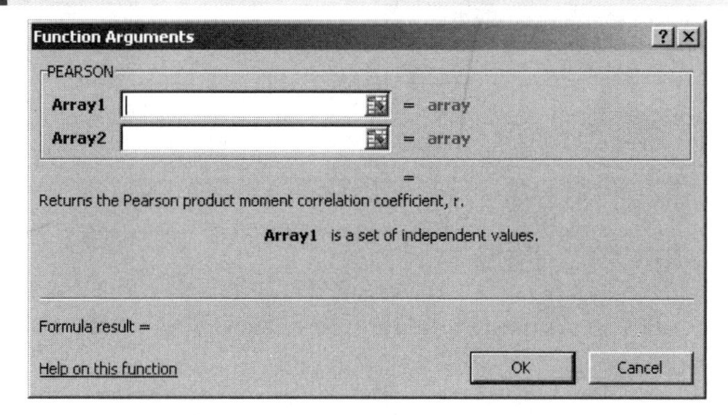

Figure 18.2 The PEARSON Function Returning the Pearson Coefficient Value

C23 | fx =PEARSON(B2:B21,C2:C21)

	A	B	C	D	E
1	ID	Income	Level of Education		
2	1	$ 59,602	15		
3	2	$ 57,108	11		
4	3	$ 42,027	14		
5	4	$ 58,404	12		
6	5	$ 66,879	10		
7	6	$ 34,054	12		
8	7	$ 51,579	16		
9	8	$ 34,123	12		
10	9	$ 35,052	14		
11	10	$ 35,976	13		
12	11	$ 57,250	16		
13	12	$ 22,134	10		
14	13	$ 39,274	16		
15	14	$ 65,789	18		
16	15	$ 46,775	13		
17	16	$ 30,552	14		
18	17	$ 76,897	16		
19	18	$ 88,767	16		
20	19	$ 65,789	16		
21	20	$ 85,645	18		
22					
23	PEARSON		0.53		

Check Your Understanding

To check your understanding of the PEARSON function, do the following two problems and check your answers in Appendix A.

QS 18a. Compute the Pearson coefficient value for 10 classrooms between years in teaching and teaching evaluations from high school students, with the latter measured from 1 to 25 with 25 being a terrific teacher. Use Data Set 18a.

QS 18b. Compute the Pearson coefficient for 15 communities between consumption of ice cream per capita in quarts and crime rate, with the latter measured by the CR scale from 1 to 10 with 10 being high crime. Use Data Set 18b.

EXCEL QUICKGUIDE 19

The INTERCEPT Function

What the INTERCEPT Function Does

The INTERCEPT function computes the intercept value, the point at which the regression line crosses the *y*-axis.

The Data Set

The data set used in this example is titled INTERCEPT, and the question is, "What is the intercept for the regression line for wins predicted by injuries for 15 teams?"

Variable	*Description*
Wins (*Y*)	Number of wins last season
Injuries (*X*)	Average number of weekly injuries

Using the INTERCEPT Function

1. Click on the cell where you want the INTERCEPT function to be placed. (In the data set, the cell is C18.)
2. Click the Formulas tab → Insert Function button (*fx*) and locate and double-click on the INTERCEPT function. You will see the Function Arguments dialog box, as shown in Figure 19.1.
3. Click the RefEdit button in the Known_y's entry box and drag the mouse over the range of cells (C2 through C16) you want included in the analysis. Click the RefEdit button.
4. Repeat Step 3 for Known_x's (Cells C2 through C16) and then click the RefEdit button
5. Click OK. The INTERCEPT function returns a value of 4.52 in Cell C18, as you see in Figure 19.2. You can see the syntax for the function in the formula bar at the top of the worksheet.

Related Functions: COVARIANCE.S, CORREL, PEARSON, SLOPE, TREND, FORECAST, RSQ

Figure 19.1 The INTERCEPT Function Arguments Dialog Box

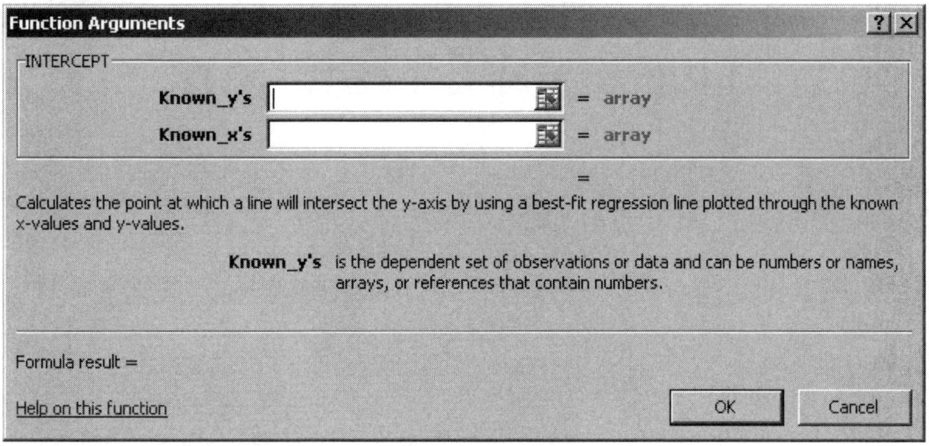

Function Arguments

INTERCEPT

Known_y's = array

Known_x's = array

=

Calculates the point at which a line will intersect the y-axis by using a best-fit regression line plotted through the known x-values and y-values.

Known_y's is the dependent set of observations or data and can be numbers or names, arrays, or references that contain numbers.

Formula result =

Help on this function

OK Cancel

Figure 19.2 The INTERCEPT Function Returning the Value of the y-Intercept

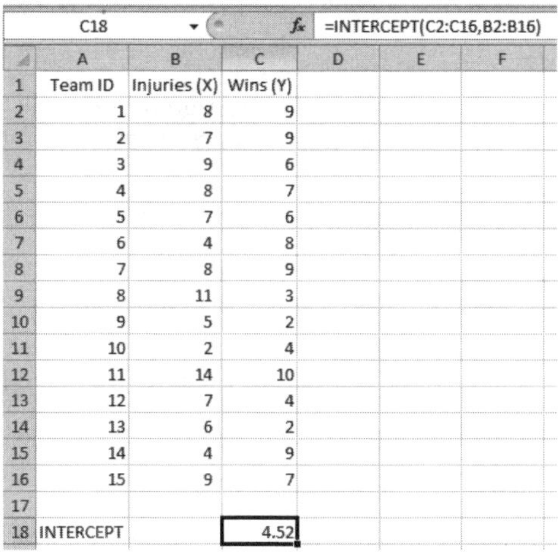

C18 =INTERCEPT(C2:C16,B2:B16)

	A	B	C	D	E	F
1	Team ID	Injuries (X)	Wins (Y)			
2	1	8	9			
3	2	7	9			
4	3	9	6			
5	4	8	7			
6	5	7	6			
7	6	4	8			
8	7	8	9			
9	8	11	3			
10	9	5	2			
11	10	2	4			
12	11	14	10			
13	12	7	4			
14	13	6	2			
15	14	4	9			
16	15	9	7			
17						
18	INTERCEPT		4.52			

Check Your Understanding

To check your understanding of the INTERCEPT function, do the following two problems and check your answers in Appendix A.

QS 19a. Compute the intercept for the regression line predicting future income from grade point average for 20 accounting students. Use Data Set 19a.

QS 19b. Compute the intercept for the regression line predicting success (on a scale of 1 to 10 with 10 being most successful) from years of experience. Use Data Set 19b.

EXCEL QUICKGUIDE 20

The SLOPE Function

What the SLOPE Function Does

The SLOPE function computes the slope of the regression line.

The Data Set

The data set used in this example is titled SLOPE, and the question is, "What is the slope of the regression line that predicts wins from injuries?"

Variable	*Description*
Wins (*Y*)	Number of wins last season
Injuries (*X*)	Average number of weekly injuries

Using the SLOPE Function

1. Highlight the cell where you want the value of the SLOPE function to be returned. (In the data set, it is C18.)
2. Click the Formulas tab → Insert Function button (*fx*) and locate and double-click on the SLOPE function. You will see the Function Arguments dialog box, as shown in Figure 20.1.
3. Click the RefEdit button in the Known_y's entry box and drag the mouse over the range of cells (C2 through C16) you want included in the analysis. Click the RefEdit button.
4. Repeat Step 3 for Known_x's (Cells B2 through B16) and then click the RefEdit button.
5. Click OK. The SLOPE function returns its value as 0.25 in Cell C18, as you see in Figure 20.2. You can see the syntax for the function in the formula bar at the top of the worksheet.

Related Functions: COVARIANCE.S, CORREL, PEARSON, INTERCEPT, TREND, FORECAST, RSQ

Figure 20.1 The SLOPE Function Arguments Dialog Box

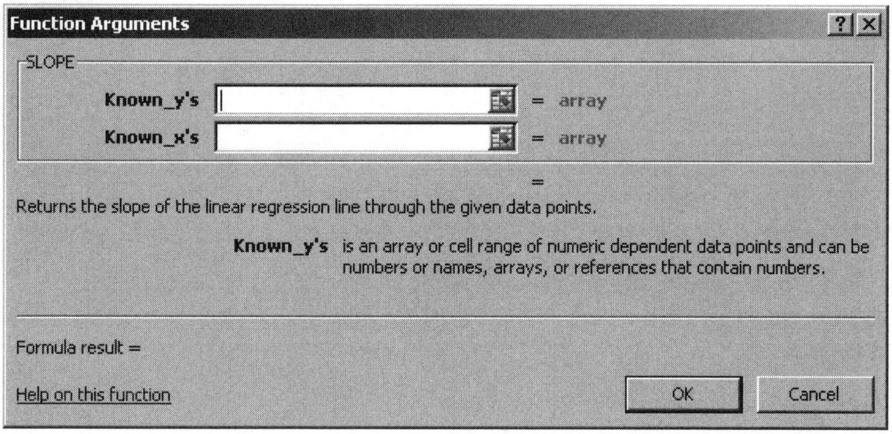

Function Arguments

SLOPE

Known_y's = array

Known_x's = array

=

Returns the slope of the linear regression line through the given data points.

Known_y's is an array or cell range of numeric dependent data points and can be numbers or names, arrays, or references that contain numbers.

Formula result =

Help on this function

OK Cancel

Figure 20.2 The SLOPE Function Returning the Slope of the Regression Line

C18 f_x =SLOPE(C2:C16,B2:B16)

	A	B	C	D	E	F
1	Team ID	Injuries (X)	Wins (Y)			
2	1	8	9			
3	2	7	9			
4	3	9	6			
5	4	8	7			
6	5	7	6			
7	6	4	8			
8	7	8	9			
9	8	11	3			
10	9	5	2			
11	10	2	4			
12	11	14	10			
13	12	7	4			
14	13	6	2			
15	14	4	9			
16	15	9	7			
17						
18	SLOPE		0.25			

Check Your Understanding

To check your understanding of the SLOPE function, do the following two problems and check your answers in Appendix A.

QS 20a. Compute the slope for the regression line predicting future income from grade point average for 20 accounting students. Use Data Set 20a.

QS 20b. Compute the slope for the regression line predicting success (on a scale of 1 to 10 with 10 being most successful) from years of experience for 10 employees. Use Data Set 20b.

EXCEL QUICKGUIDE 21

The TREND Function

What the TREND Function Does

The TREND function uses the regression line values to predict outcomes.

The Data Set

The data set used in this example is titled TREND, and the question is, "What is the predicted level of wins given injuries for members of 15 volleyball teams?"

Variable	*Description*
Injuries (*X*)	Average number of player injuries
Wins (*Y*)	Season wins

Using the TREND Function

1. Highlight the cells where you want the array of TREND values to appear. (In the data set, the cells are E2 through E4.)

2. Click the Formulas tab → Insert Function button (*fx*) and locate and double-click on the TREND function. You will see the Function Arguments dialog box, as shown in Figure 21.1.

3. Click the RefEdit button in the Known_y's entry box and drag the mouse over the range of cells (C2 through C16) you want included in the analysis. Click the RefEdit button.

4. Click the RefEdit button for Known_x's (Cells B2 through B16) and then click the RefEdit button.

5. Click the RefEdit button for New_x's (Cells D2 through D4) and then click the RefEdit button.

6. Because this is an array, use the Ctrl+Shift+Enter key combination to produce the TREND function and three predicted scores of 5.27, 4.77, and 7.02, as you see in Figure 21.2. You can see the syntax for the function in the formula bar at the top of the worksheet.

Related Functions: COVARIANCE.S, CORREL, PEARSON, INTERCEPT, SLOPE, FORECAST, RSQ

Figure 21.1 The TREND Function Arguments Dialog Box

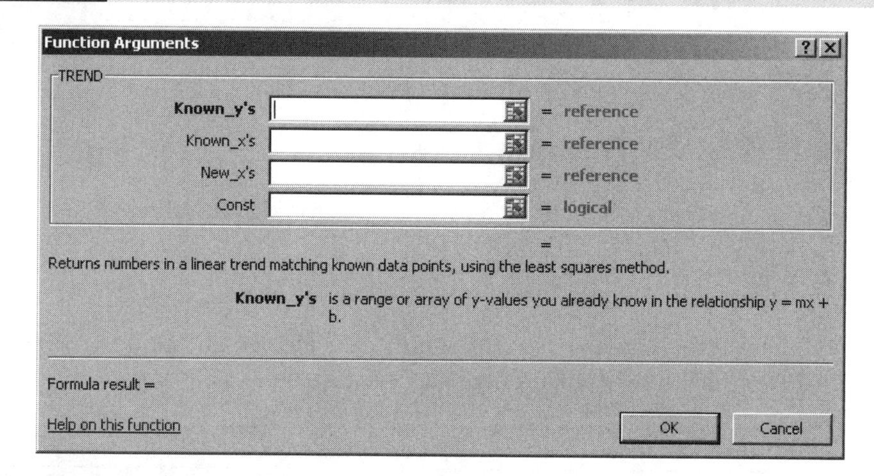

Figure 21.2 The TREND Function Returning the Expected *Y* Values of the Regression Line for Given *X*'s

E2 | {=TREND(C2:C16,B2:B16,D2:D4)}

	A	B	C	D	E	F
1	Team ID	Injuries (X)	Wins (Y)	New Injuries	TREND	
2	1	8	9	3	5.27	
3	2	7	9	1	4.77	
4	3	9	6	10	7.02	
5	4	8	7			
6	5	7	6			
7	6	4	8			
8	7	8	9			
9	8	11	3			
10	9	5	2			
11	10	2	4			
12	11	14	10			
13	12	7	4			
14	13	6	2			
15	14	4	9			
16	15	9	7			

Check Your Understanding

To check your understanding of the TREND function, do the following two problems and check your answers in Appendix A.

QS 21a. Compute the trend for new income scores. Use Data Set 21a.

QS 21b. Compute the trend for new success scores. Use Data Set 21b.

EXCEL QUICKGUIDE 22

The FORECAST Function

What the FORECAST Function Does

The FORECAST function computes a predicted value for known values of *X*.

The Data Set

The data set used in this example is titled FORECAST, and the question is, "What is the predicted GPA for nine newly ranked high school students?"

Variable	*Description*
GPA (*Y*)	Grade point average
Rank (*X*)	High school rank (from 1 to 5)

Using the FORECAST Function

1. Highlight the cells where you want the array of FORECAST values to appear. (In the data set, the cells are E2 through E10.)
2. Click the Formulas tab → Insert Function button (*fx*) and locate and double-click on the FORECAST function. You will see the Function Arguments dialog box, as shown in Figure 2.1.
3. Click the RefEdit button in the X entry box and drag the mouse over the range of cells (D2 through D10) you are using to predict the Y. Click the RefEdit button.
4. Click the RefEdit button in the Known_y's entry box and drag the mouse over the range of cells (B2 through B26) you want included in the analysis. Click the RefEdit button.
5. Click the RefEdit button in the Known_x's entry box and drag the mouse over the range of cells (C2 through C26) you want included in the analysis. Click the RefEdit button.
6. Because this is an array, press the Ctrl+Shift+Enter key combination to produce the FORECAST function and the forecast scores of 3.68, 3.32, and so on, as you see in Figure 22.2. You can see the syntax for the function in the formula bar at the top of the worksheet.

Related Functions: COVARIANCE.S, CORREL, PEARSON, INTERCEPT, SLOPE, TREND, RSQ

Figure 22.1 The FORECAST Function Arguments Dialog Box

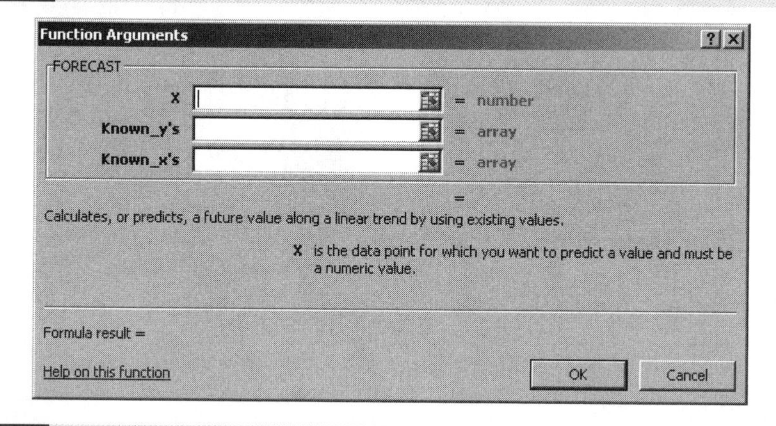

Figure 22.2 The FORECAST Function Returning the Expected *Y* Values of the Regression Line for Given *X*s

E2 | *fx* {=FORECAST(D2:D10,B2:B26,C2:C26)}

	A	B	C	D	E	F	G
1	ID	GPA (Y)	RANK (X)	New Rank	FORECAST		
2	1	4.0	5	5	3.68		
3	2	3.2	4	4	3.32		
4	3	3.9	5	5	3.68		
5	4	2.7	3	3	2.97		
6	5	3.6	5	4	3.32		
7	6	1.9	1	5	3.68		
8	7	2.4	2	6	4.04		
9	8	3.6	4	5	3.68		
10	9	3.7	4	4	3.32		
11	10	3.2	2				
12	11	2.7	3				
13	12	2.1	1				
14	13	1.9	3				
15	14	1.7	3				
16	15	2.9	3				
17	16	3.1	2				
18	17	2.7	3				
19	18	3.6	4				
20	19	2.6	3				
21	20	3.1	4				
22	21	2.6	1				
23	22	2.9	1				
24	23	4.0	4				
25	24	3.8	3				
26	25	3.0	4				

Check Your Understanding

To check your understanding of the FORECAST function, do the following two problems and check your answers in Appendix A.

QS 22a. Forecast number of new losses from previous 10 years of win-loss data for a large eastern university football team. Use Data Set 22a.

QS 22b. Forecast new American history scores from previous world history test scores. Use Data Set 22b.

EXCEL QUICKGUIDE 23

The RSQ Function

What the RSQ Function Does

The RSQ function computes the square of the product-moment correlation between two variables.

The Data Set

The data set used in this example is, titled RSQ, and the question is, "What is the RSQ, or the amount of variance accounted for, in the relationship between income and level of education?"

Variable	*Description*
Income	Annual income in dollars
Level of Education	Years of education

Using the RSQ Function

1. Click on the cell where you want the RSQ function to be placed. (In the data set, the cell is C23.)
2. Click the Formulas tab → Insert Function button (*fx*) and locate and double-click on the RSQ function. You will see the Function Arguments dialog box, as shown in Figure 23.1.
3. Click the RefEdit button in the Known_y's entry box and drag the mouse over the range of cells (B2 through B21) you want included in the analysis. Click the RefEdit button.
4. Repeat Step 3 for Known_x's (Cells C2 through C21) and then click the RefEdit button.
5. Click OK. The RSQ function returns the value .28 (rounded) in Cell C23, as you see in Figure 23.2. You can see the syntax for the function in the formula bar at the top of the worksheet.

Related Functions: COVARIANCE.S, CORREL, PEARSON, INTERCEPT, SLOPE, TREND, FORECAST

Figure 23.1 The RSQ Function Arguments Dialog Box

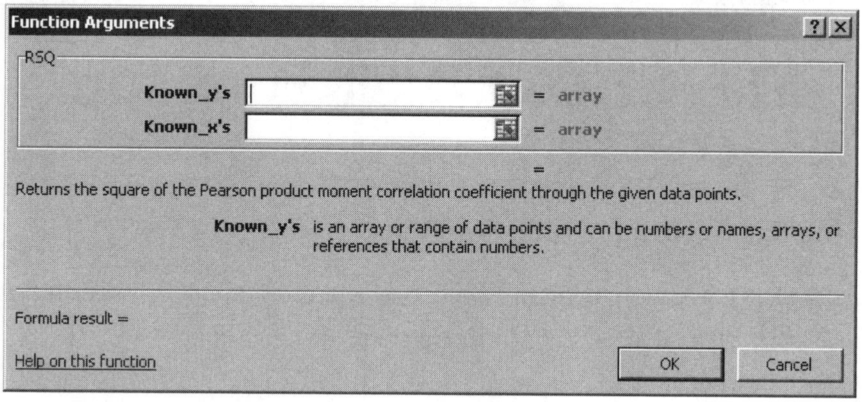

Function Arguments

RSQ

Known_y's = array

Known_x's = array

=

Returns the square of the Pearson product moment correlation coefficient through the given data points.

Known_y's is an array or range of data points and can be numbers or names, arrays, or references that contain numbers.

Formula result =

Help on this function

OK Cancel

Figure 23.2 The RSQ Function Returning the Squared Correlation

C23 =RSQ(B2:B21,C2:C21)

	A	B	C	D
1	ID	Income	Level of Education	
2	1	$ 59,602	15	
3	2	$ 57,108	11	
4	3	$ 42,027	14	
5	4	$ 58,404	12	
6	5	$ 66,879	10	
7	6	$ 34,054	12	
8	7	$ 51,579	16	
9	8	$ 34,123	12	
10	9	$ 35,052	14	
11	10	$ 35,976	13	
12	11	$ 57,250	16	
13	12	$ 22,134	10	
14	13	$ 39,274	16	
15	14	$ 65,789	18	
16	15	$ 46,775	13	
17	16	$ 30,552	14	
18	17	$ 76,897	16	
19	18	$ 88,767	16	
20	19	$ 65,789	16	
21	20	$ 85,645	18	
22				
23	RSQ		0.28	

Check Your Understanding

To check your understanding of the RSQ function, do the following two problems and check your answers in Appendix A.

QS 23a. Compute the RSQ between height in inches and weight for 20 first-year college students. Use Data Set 23a.

QS 23b. Compute the RSQ between quality of health care (from 1 to 5 with 5 being most healthy) and decline in infant mortality (deaths per 1,000 live-born infants) for five communities. Use Data Set 23b.

EXCEL QUICKGUIDE 24

The CHISQ.DIST Function

What the CHISQ.DIST Function Does

The CHISQ.DIST function computes the probability of a value associated with a chi-square (χ^2) value. (Use the CHISQ.TEST function for computing the χ^2 value.)

The Data Set

The data set used in this example is titled CHISQ.DIST, and the question being asked is, "What is the probability of a type I error or alpha level associated with a chi-square of 0.29?"

Variable	*Value*
Chi Square Value	Value of chi-square
Degrees of Freedom	Degrees of freedom associated with the original chi-square analysis

Using the CHISQ.DIST Function

1. Click on the cell where you want the CHISQ.DIST function to be placed. (In the data set, the cell is B4.)
2. Click the Formulas tab → Insert Function button (*fx*) and locate and double-click on the CHISQ.DIST function. You will see the Function Arguments dialog box, as shown in Figure 24.1.
3. Click the RefEdit button in the X entry box and enter the chi-square value. In this example, it is Cell B1. Click the RefEdit button.
4. Click the RefEdit button in the Deg_freedom entry box and click on the cell (B2) indicating the degrees of freedom. Click the RefEdit button.
5. Enter True for the Cumulative value.
6. Click OK, and the CHISQ.DIST function returns a value of .13 in Cell B4, as you see in Figure 24.2, indicating that it is highly likely that this value occurred by

chance. Note that you can see the syntax for the function in the formula bar at the top of the worksheet.

Related Functions: CHISQ.TEST

Figure 24.1 The CHISQ.DIST Function Arguments Dialog Box

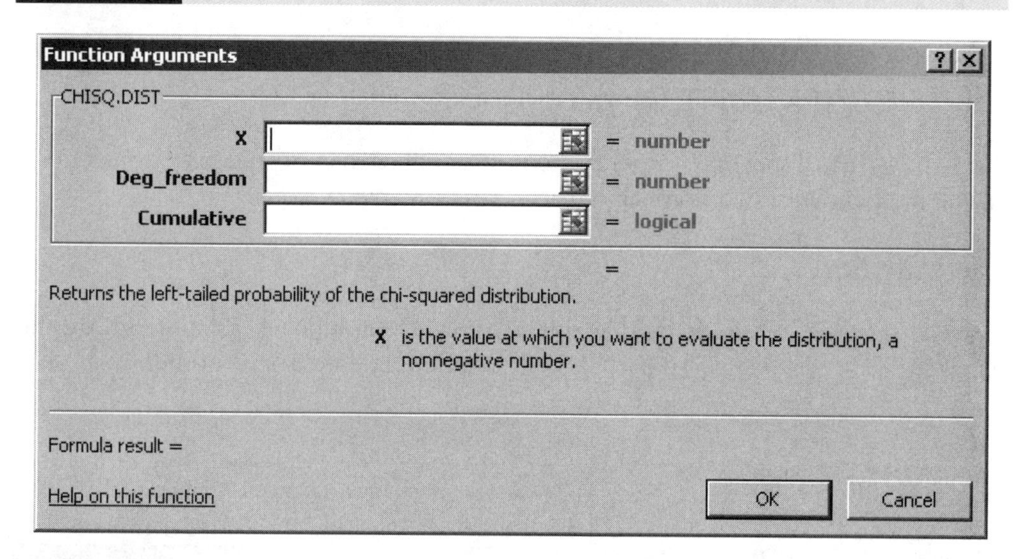

Figure 24.2 The CHISQ.DIST Function Returning the Probability Associated With a Chi-Square Value

B4 | f_x =CHISQ.DIST(B1,B2,TRUE)

	A	B	C	D	E
1	Chi Square Value	0.29			
2	Degrees of Freedom	2			
3					
4	CHISQ.DIST	0.13			

Check Your Understanding

To check your understanding of the CHISQ.DIST function, do the following two problems and check your answers in Appendix A.

QS 24a. Compute the CHISQ.DIST value for a chi-square value of 7.8 with 10 degrees of freedom. Use Data Set 24a.

QS 24b. Compute the CHISQ.DIST value for a chi-square value of 2.1 with 4 degrees of freedom. Use Data Set 24b.

EXCEL QUICKGUIDE 25

The CHISQ.TEST Function

What the CHISQ.TEST Function Does

The CHISQ.TEST function computes the probability of the chi-square (χ^2) value for a test of independence of a nominal or categorical variable.

The Data Set

The data set used in this example is titled CHISQ.TEST, and the question is, "Are the actual and expected values for party affiliation independent of one another?"

Variable	*Description*
Party Affiliation	Frequency of actual and expected values for Democratic, Republican, and independent voters

Using the CHISQ.TEST Function

1. Click on the cell where you want the CHISQ.TEST function returned. (In the data set, the cell is B10.)
2. Click the Formulas tab → Insert Function button (*fx*) and locate and double-click on the CHISQ.TEST function. You will see the Function Arguments dialog box, as shown in Figure 25.1.
3. Click the RefEdit button in the Actual_range entry box and drag the mouse over the range of cells (B8 through D8) you want included in the analysis. Click the RefEdit button.
4. Click the RefEdit button in the Expected_range entry box and drag the mouse over the range of cells (B4 through D4) you want included in the analysis.
5. Click the RefEdit button or press the Enter key.
6. Click OK. The CHISQ.TEST function returns its value of .29 in Cell B10, as you see in Figure 25.2. Note that you can see the syntax for the function in the formula bar at the top of the worksheet.

Related Functions: CHISQ.DIST

Figure 25.1 The CHISQ.TEST Function Arguments Dialog Box

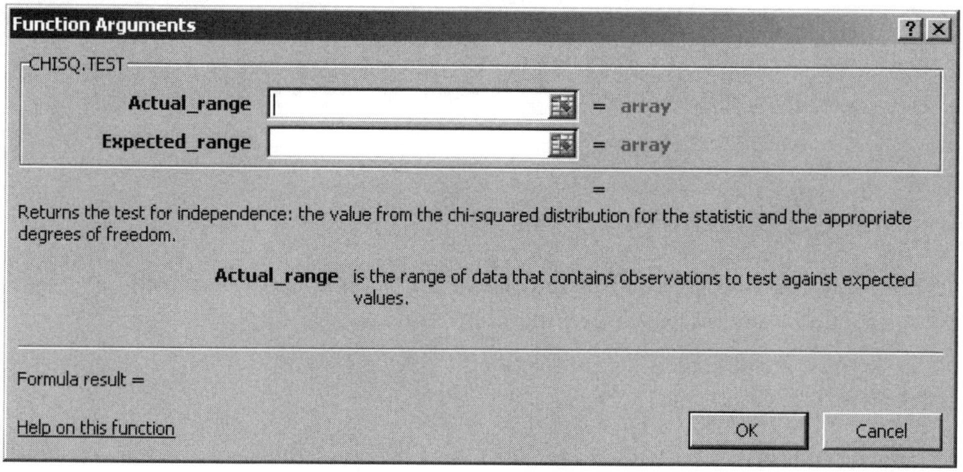

Function Arguments

CHISQ.TEST

Actual_range = array

Expected_range = array

=

Returns the test for independence: the value from the chi-squared distribution for the statistic and the appropriate degrees of freedom.

Actual_range is the range of data that contains observations to test against expected values.

Formula result =

Help on this function

OK Cancel

Figure 25.2 The CHISQ.TEST Function Returning the Probability of the Chi-Square (χ^2) Value

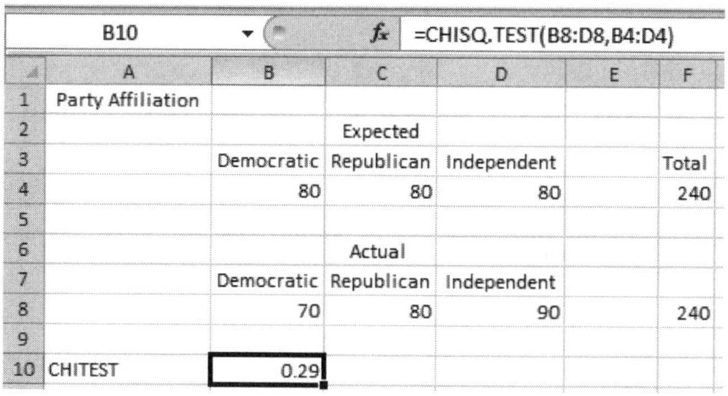

B10 f_x =CHISQ.TEST(B8:D8,B4:D4)

	A	B	C	D	E	F
1	Party Affiliation					
2			Expected			
3		Democratic	Republican	Independent		Total
4		80	80	80		240
5						
6			Actual			
7		Democratic	Republican	Independent		
8		70	80	90		240
9						
10	CHITEST	0.29				

Check Your Understanding

To check your understanding of the CHISQ.TEST function, do the following two problems and check your answers in Appendix A.

QS 25a. Compute the probability of the chi-square value for groups of expected and users of a community center. Use Data Set 25a.

QS 25b. Compute the probability of the chi-square value for groups of users of three different brands of detergent. Use Data Set 25b.

EXCEL QUICKGUIDE 26

The F.DIST Function

What the F.DIST Function Does

The F.DIST function computes the probability of a value associated with an *F* value. (Use the F.TEST function for computing the *F* value.)

The Data Set

The data set in this example is titled F.DIST, and the question being asked is, "What is the probability of a type I error or alpha level associated with an *F* value of 1.79 with 2 and 35 degrees of freedom?"

Variable	*Value*
F	Value of *F*
Degrees of Freedom (Numerator)	Degrees of freedom associated with the numerator
Degrees of Freedom (Denominator)	Degrees of freedom associated with the denominator

Using the F.DIST Function

1. Click on the cell where you want the F.DIST function to be placed. (In the data set, the cell is B5.)
2. Click the Formulas tab → Insert Function button (*fx*) and locate and double-click on the F.DIST function. You will see the Function Arguments dialog box, as shown in Figure 26.1.
3. Click the RefEdit button in the X entry box, enter the *F* value (Cell B1), and click the RefEdit button.
4. Click the RefEdit button in the Deg_freedom1 entry box and enter the degrees of freedom value of the numerator (Cell B2). Click the RefEdit button.
5. Repeat Step 4 for the Deg_freedom2 entry box, entering the degrees of freedom of the denominator (Cell B3). Click the RefEdit button.
6. Enter True in the Cumulative box.

7. Click OK. The F.DIST function returns a value of .82 in Cell B5, as you see in Figure 26.2, indicating that the *F* value is significant beyond the .05 level. Note that you can see the syntax for the function in the formula bar at the top of the worksheet.

Related Functions: F.TEST, T.DIST, T.TEST, Z.TEST

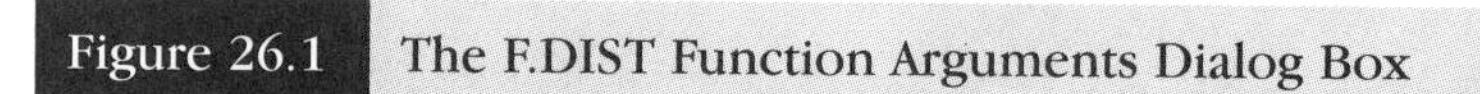

Figure 26.1 The F.DIST Function Arguments Dialog Box

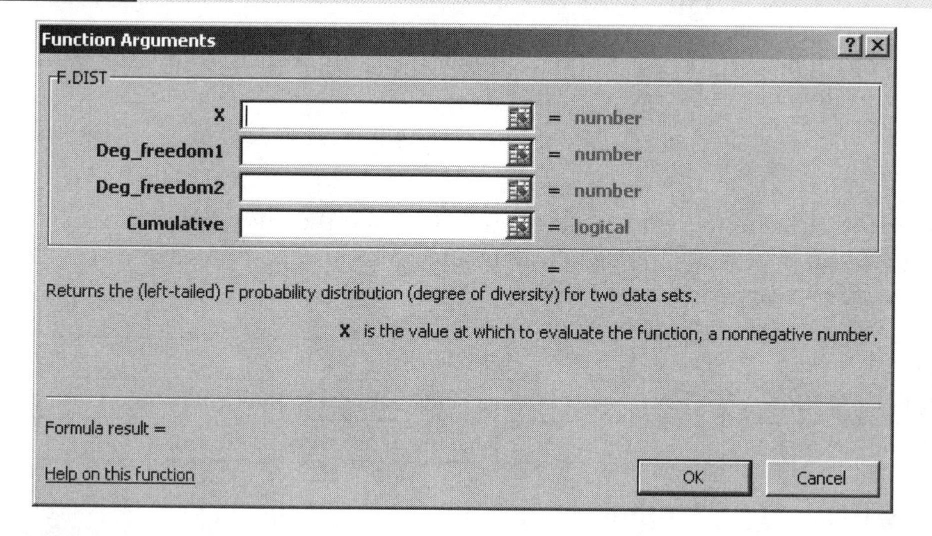

Figure 26.2 The F.DIST Function Returning the Probability Associated With an *F* Value

B5 | *fx* =F.DIST(B1,B2,B3,TRUE)

	A	B	C	D
1	F Value	1.79		
2	Degrees of Freedom (1 - Numerator)	2		
3	Degrees of Freedom (2 - Denominator)	35		
4				
5	F.DIST	0.82		

Check Your Understanding

To check your understanding of the F.DIST function, do the following two problems and check your answers in Appendix A.

QS 26a. Compute the F.DIST value for the data shown in Data Set 26a.

QS 26b. Compute the F.DIST value for the data shown in Data Set 26b.

EXCEL QUICKGUIDE 27

The F.TEST Function

What the F.TEST Function Does

The F.TEST function computes the probability that the associated *F* value is not significantly different from zero.

The Data Set

The data set used in this example is titled F.TEST, and the question is, "Does studying using an iPad as a study aid significantly affect the final test score of two different groups of first-year college students?"

Variable	*Description*
iPad	Final test score with the help program
No iPad	Final test score without the help program

Using the F.TEST Function

To use the F.TEST function, follow these steps:

1. Click on the cell where you want the F.TEST function returned. (In the data set, the cell is C23.)
2. Click the Formulas tab → Insert Function button (*fx*) and locate and double-click on the F.TEST function. You will see the Function Arguments dialog box, as shown in Figure 27.1.
3. Click the RefEdit button in the Array1 entry box. Drag the mouse over the range of cells (B2 through B21) you want included in the analysis and click the RefEdit button.
4. Click the RefEdit button in the Array2 entry box. Drag the mouse over the range of cells (C2 through C21) you want included in the analysis and click the RefEdit button.
5. Click OK. The F.TEST function returns its value of .12 in Cell C23, as you see in Figure 27.2, indicating that the difference between groups on final test score is not significant at the .05 level. Note that you can see the syntax for the function in the formula bar at the top of the worksheet.

Related Functions: F.DIST, T.DIST, T.TEST, Z.TEST

Figure 27.1 The F.TEST Function Arguments Dialog Box

Function Arguments

F.TEST

Array1 = array

Array2 = array

=

Returns the result of an F-test, the two-tailed probability that the variances in Array1 and Array2 are not significantly different.

Array1 is the first array or range of data and can be numbers or names, arrays, or references that contain numbers (blanks are ignored).

Formula result =

Help on this function

OK Cancel

Figure 27.2 The F.TEST Function Returning the Probability That the *F* Value Is Different From Zero

C23 =F.TEST(B2:B21,C2:C21)

	A	B	C
1	ID	iPod	No iPod
2	1	87	77
3	2	46	56
4	3	51	45
5	4	78	67
6	5	88	89
7	6	45	78
8	7	76	66
9	8	78	78
10	9	87	98
11	10	91	98
12	11	91	94
13	12	28	76
14	13	75	78
15	14	65	71
16	15	45	68
17	16	83	89
18	17	46	81
19	18	89	70
20	19	85	91
21	20	65	77
22			
23	FTEST		0.12

Check Your Understanding

To check your understanding of the F.TEST function, do the following two problems and check your answers in Appendix A.

QS 27a. Compute the F.TEST value for the difference between attitudes toward using an iPad or not on ease of completing calculation tasks. Use Data Set 27a.

QS 27b. Compute F.TEST value for the difference between two groups of calculus test scores for two different samples of 15 college juniors. Use Data Set 27b.

EXCEL QUICKGUIDE 28

The T.DIST Function

What the T.DIST Function Does

The T.DIST function computes the probability of a value associated with the Student's *t* value. (Use the T.TEST function for computing the *t* value.)

The Data Set

The data set in this example is titled T.DIST, and the question being asked is, "What is the probability of a type I error or alpha level associated with a one-tailed test and a *t* value of 1.96 with 50 degrees of freedom?"

Variable	*Value*
X	Value of *t*
Degrees of Freedom	Degrees of freedom
Tails	1 = one-tailed test, 2 = two-tailed test

Using the T.DIST Function

1. Click on the cell where you want the T.DIST function to be placed. (In the data set, the cell is B4.)
2. Click the Formulas tab → Insert Function button (*fx*) and locate and double-click on the T.DIST function. You will see the Function Arguments dialog box, as shown in Figure 28.1.
3. Click the RefEdit button in the X entry box and enter the *t* value. In this example, it is Cell B1. Click the RefEdit button.
4. Repeat Step 3 for the Deg_freedom entry box and enter the degrees of freedom value (Cell B2). Click the RefEdit button.
5. Enter True in the Cumulative box.

6. Click OK. The T.DIST function returns a value of .972 in Cell B4, as you see in Figure 28.2, indicating that the t value is significant beyond the .05 level. Note that you can see the syntax for the function in the formula bar at the top of the worksheet.

Related Functions: F.DIST, F.TEST, T.TEST, Z.TEST

Figure 28.1 The T.DIST Function Arguments Dialog Box

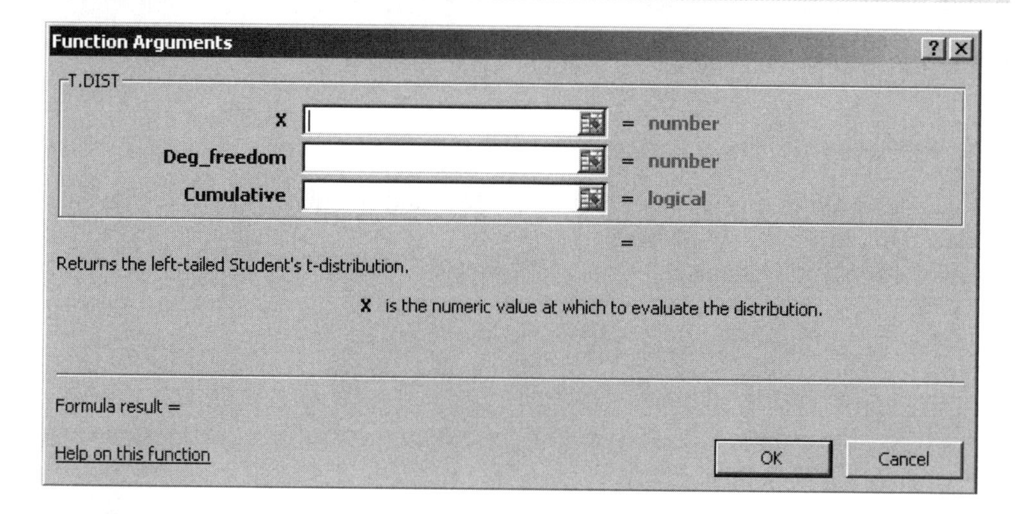

Figure 28.2 The T.DIST Function Returning the Probability Associated With a t Value

B4   fx =T.DIST(B1,B2,TRUE)

	A	B	C	D	E
1	X	1.96			
2	Degrees of Freedom	50			
3					
4	T.DIST	0.972			

Check Your Understanding

To check your understanding of the T.DIST function, do the following two problems and check your answers in Appendix A.

QS 28a. Compute the T.DIST value for the data shown in Data Set 28a.

QS 28b. Compute the T.DIST value for the data shown in Data Set 28b.

EXCEL QUICKGUIDE 29

The T.TEST Function

What the T.TEST Function Does

The T.TEST function computes the Student's *t* value.

The Data Set

The data set used in this example is titled T.TEST, and the question is, "Is there a significant difference in achievement scores between fall and spring tests of the same group of 20 college students?"

Variable	*Description*
Fall	Fall test scores
Spring	Spring test scores

Using the T.TEST Function

To use the T.TEST function, follow these steps:

1. Click on the cell where you want the T.TEST function returned. (In the data set, the cell is C23.)
2. Click the Formulas tab → Insert Function button (*fx*) and locate and double-click on the T.TEST function. You will see the Function Arguments dialog box, as shown in Figure 29.1.
3. Click the RefEdit button in the Array1 entry box. Drag the mouse over the range of cells (B2 through B21) you want included in the analysis and click the RefEdit button.
4. Repeat Step 3 for the Array2 entry box. Drag the mouse over the range of cells (C2 through C21) you want included in the analysis and click the RefEdit button.
5. In the Tails entry box, enter the value 2, and in the Type entry box, enter the value 1 (for a paired *t*-test).
6. Click OK. The T.TEST function returns a value of .14 in Cell C23, as you see in Figure 29.2, indicating that the probability of the associated *t* value occurring by chance alone is .14. Note that you can see the syntax for the function in the formula bar at the top of the worksheet.

Related Functions: F.DIST, F.TEST, T.DIST, Z.TEST

Figure 29.1 The T.TEST Function Arguments Dialog Box

Function Arguments

T.TEST

Array1 = array
Array2 = array
Tails = number
Type = number

=

Returns the probability associated with a Student's t-Test.

Array1 is the first data set.

Formula result =

Help on this function

OK Cancel

Figure 29.2 The T.TEST Function Returning the Probability of the Student's *t* Value

C23 f_x =T.TEST(B2:B21,C2:C21,2,1)

	A	B	C	D	E	F
1	ID	Fall	Spring			
2	6	5	4			
3	2	5	5			
4	3	4	4			
5	4	7	3			
6	5	8	3			
7	6	7	8			
8	7	6	7			
9	8	3	6			
10	9	6	5			
11	10	2	6			
12	11	6	9			
13	12	4	8			
14	13	5	9			
15	14	7	8			
16	15	8	9			
17	16	7	8			
18	17	5	9			
19	18	9	8			
20	19	8	9			
21	20	6	7			
22						
23	T.TEST		0.14			

Check Your Understanding

To check your understanding of the T.TEST function, do the following two problems and check your answers in Appendix A.

QS 29a. Compute the probability associated with a *t*-test for the difference in effectiveness between two methods of training how to speak a foreign language as measured by a test with scores ranging from 0 to 100 given to 25 English-speaking children in elementary school. Use Data Set 29a.

QS 29b. Compute the probability associated with a *t*-test for the difference in quality between two different dishes cooked by 10 chefs as measured on a 5-point scale with 5 being excellent as judged by expert chefs. Use Data Set 29b.

EXCEL QUICKGUIDE 30

The Z.TEST Function

What the Z.TEST Function Does

The Z.TEST function computes the probability that a single score belongs to a population of scores.

The Data Set

The data set used in this example is titled Z.TEST, and the question is, "Does a score of 82 belong to the population set of scores?"

Variable	*Description*
Memo	Score on a recall memory test ranging from 0 to 100
X	The score to be tested
Sigma	The population standard deviation (computed using the STDEV.P function)

Using the Z.TEST Function

To use the Z.TEST function, follow these steps:

1. Click on the cell where you want the Z.TEST function returned. (In the data set, the cell is B18.)
2. Click the Formulas tab → Insert Function button (*fx*) and locate and double-click on the Z.TEST function. You will see the Function Arguments dialog box, as shown in Figure 30.1.
3. Click the RefEdit button in the Array entry box. Drag the mouse over the range of cells (B2 through B16) you want included in the analysis and click the RefEdit button.
4. Click the RefEdit button in the X entry box and click on the X value for which you want to compute the *Z*-test (Cell C2). Click the RefEdit button.
5. Click the RefEdit button, click on the population standard deviation (Cell D2), and click the RefEdit button.

6. Click OK. The Z.TEST function returns a value of .94 in Cell B18, as you see in Figure 30.2, indicating that it is highly likely the *X* value belongs to the set of sample scores and is not unique. Note that you can see the syntax for the function in the formula bar at the top of the worksheet.

Related Functions: F.DIST, F.TEST, T.DIST, T.TEST

Figure 30.1 The Z.TEST Function Arguments Dialog Box

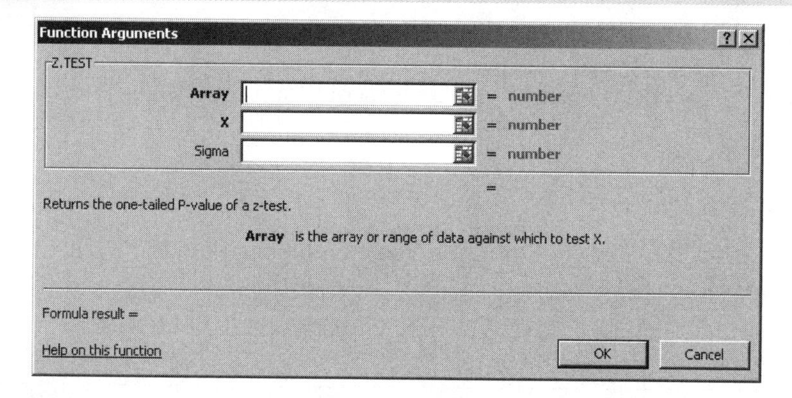

Figure 30.2 The Z.TEST Function Returning the Probability That a Single Value Belongs to a Population

B18 =Z.TEST(B2:B16,C2,D2)

	A	B	C	D	E
1	ID	MEMO	X	Sigma	
2	1	67	82	11.44	
3	2	67			
4	3	87			
5	4	89			
6	5	87			
7	6	77			
8	7	73			
9	8	74			
10	9	68			
11	10	72			
12	11	58			
13	12	98			
14	13	98			
15	14	70			
16	15	77			
17					
18	Z.TEST	0.94			

Check Your Understanding

To check your understanding of the Z.TEST function, do the following two problems and check your answers in Appendix A.

QS 30a. Compute the probability that a math test score of 83 is characteristic of a set of test scores. Use Data Set 30a.

QS 30b. Compute the probability that a free-throw shooting percentage of 63 is unique among conference teams. Use Data Set 30b.

EXCEL QUICKGUIDE 31

The AVERAGEIF Function

What the AVERAGEIF Function Does

The AVERAGEIF function returns the average score for a set of cells based on a specific criterion.

The Data Set

The data set is titled AVERAGEIF, and the question being asked is, "What is the average spelling score for males?"

Variable	*Value*
Gender	Male = 1, Female = 2
Test Score	Score on a 20-item spelling test

Using the AVERAGEIF Function

1. Click on the cell where you want the AVERAGEIF function to be placed. (In the data set, the cell is B28.)
2. Click the Formulas tab → Insert Function button (*fx*) and locate and double-click on the AVERAGEIF function. You will see the Function Arguments dialog box, as shown in Figure 31.1.
3. Click the RefEdit button in the Range entry box and enter the range of the cells that contains the criterion. In this example, it is A2:A26. Click the RefEdit button.
4. In the Criteria box, enter the selection criterion, which in this case is 1 (for male).
5. Click the RefEdit button and enter the Average_range of the cells for which you want to compute the average (B2:B26). Click the RefEdit button.
6. Click OK. The AVERAGEIF function returns a value of 14.73 in Cell B28, as you see in Figure 31.2, indicating that the average score for males on the 20-item spelling test is 14.73. Note that you can see the syntax for the function in the formula bar at the top of the worksheet.

Related Functions: COUNT, COUNTA, COUNTBLANK, COUNTIF

Figure 31.1 The AVERAGEIF Function Arguments Dialog Box

Function Arguments

AVERAGEIF

Range = reference

Criteria = any

Average_range = reference

=

Finds average(arithmetic mean) for the cells specified by a given condition or criteria.

Range is the range of cells you want evaluated.

Formula result =

Help on this function

OK Cancel

Figure 31.2 The AVERAGEIF Function Returning the Average of Males' Scores

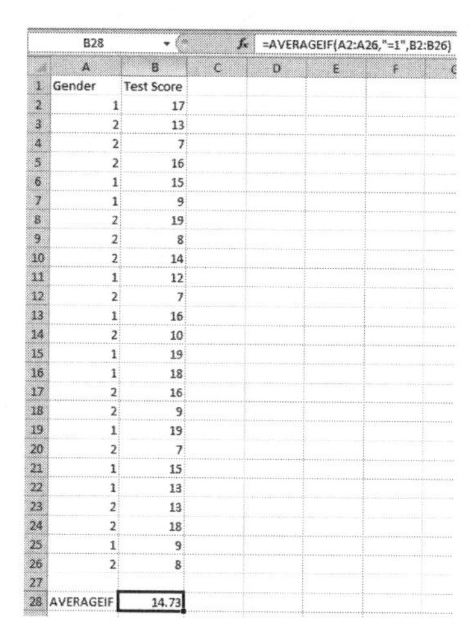

B28 =AVERAGEIF(A2:A26,"=1",B2:B26)

	A	B
1	Gender	Test Score
2	1	17
3	2	13
4	2	7
5	2	16
6	1	15
7	1	9
8	2	19
9	2	8
10	2	14
11	1	12
12	2	7
13	1	16
14	2	10
15	1	19
16	1	18
17	2	16
18	2	9
19	1	19
20	2	7
21	1	15
22	1	13
23	2	13
24	2	18
25	1	9
26	2	8
27		
28	AVERAGEIF	14.73

Check Your Understanding

To check your understanding of the AVERAGEIF function, do the following two problems and check your answers in Appendix A.

QS 31a. Compute the AVERAGEIF sales value for those members of the A Team (denoted with 1). Use Data Set 31a.

QS 31b. Compute the AVERAGEIF incidence for infection rate per 1,000 patients in hospitals in communities with fewer than 200,000 people. Use Data Set 31b.

EXCEL QUICKGUIDE 32

The COUNT Function

What the COUNT Function Does

The COUNT function returns the number of cells that contain values.

The Data Set

The data set is titled COUNT, and the question being asked is, "How many phone calls resulted in a response?"

Variable	*Value*
Response	1 or blank

Using the COUNT Function

1. Click on the cell where you want the COUNT function to be placed. (In the data set, the cell is B30.)

2. Click the Formulas tab → Insert Function button (*fx*) and locate and double-click on the COUNT function. You will see the Function Arguments dialog box, as shown in Figure 32.1.

3. Click the RefEdit button in the Value1 entry box and enter the range of the cells that contains the values you want to count. In this example, the range is B2:B28.

4. Click the OK button. The COUNT function returns a value of 18 in Cell B30, as you see in Figure 32.2, indicating that 18 of the 28 participants responded. Note that you can see the syntax for the function in the formula bar at the top of the worksheet.

Related Functions: AVERAGEIF, COUNTA, COUNTBLANK, COUNTIF

Figure 32.1 The COUNT Function Arguments Dialog Box

Function Arguments

COUNT

Value1 B2:B28 = {1;1;0;1;0;1;1;1;0;1;0;1;0;1;1;1;1;...

Value2 = number

= 18

Counts the number of cells in a range that contain numbers.

Value1: value1,value2,... are 1 to 255 arguments that can contain or refer to a variety of different types of data, but only numbers are counted.

Formula result = 18

Help on this function

OK Cancel

Figure 32.2 The COUNT Function Returning the Number of Respondents

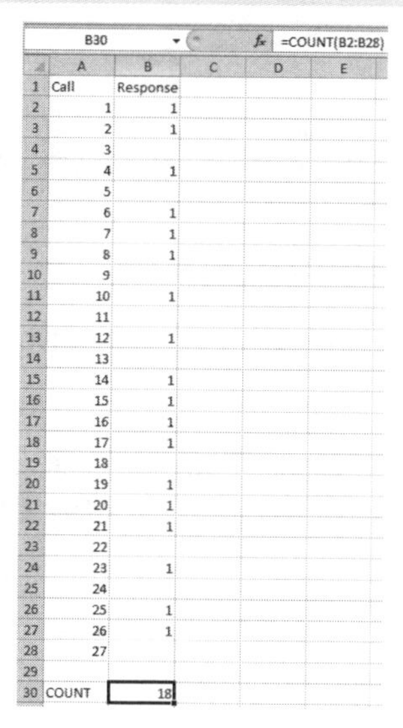

B30 =COUNT(B2:B28)

	A	B	C	D	E
1	Call	Response			
2	1	1			
3	2	1			
4	3				
5	4	1			
6	5				
7	6	1			
8	7	1			
9	8	1			
10	9				
11	10	1			
12	11				
13	12	1			
14	13				
15	14	1			
16	15	1			
17	16	1			
18	17	1			
19	18				
20	19	1			
21	20	1			
22	21	1			
23	22				
24	23	1			
25	24				
26	25	1			
27	26	1			
28	27				
29					
30	COUNT	18			

Check Your Understanding

To check your understanding of the COUNT function, do the following two problems and check your answers in Appendix A.

QS 32a. Compute the COUNT value for the number of people who purchased insurance. Use Data Set 32a.

QS 32b. Compute the COUNT value for the respondents who gave their age and those who gave their years in service. Use Data Set 32b.

EXCEL QUICKGUIDE 33

The COUNTA Function

What the COUNTA Function Does

The COUNTA function returns the number of cells that contain values, including zeros.

The Data Set

The data set is titled COUNTA, and the question being asked is, "For today's sales, how many customers purchased either beans or mac 'n' cheese?"

Variable	*Value*
Preferred	beans, mac 'n' cheese

Using the COUNTA Function

1. Click on the cell where you want the COUNTA function to be placed. (In the data set, the cell is B28.)
2. Click the Formulas tab → Insert Function button (*fx*) and locate and double-click on the COUNTA function. You will see the Function Arguments dialog box, as shown in Figure 33.1.
3. Click the RefEdit button in the Value1 entry box and enter the range of the cells that contains the values you want to count. In this example, the range is B2:B26. Click the RefEdit button.
4. Click the OK button. The COUNTA function returns a value of 14 in Cell B28, as you see in Figure 33.2, indicating 14 of the 25 customers purchased one of the two items. Note that you can see the syntax for the function in the formula bar at the top of the worksheet.

Related Functions: AVERAGEIF, COUNT, COUNTBLANK, COUNTIF

Figure 33.1 The COUNTA Function Arguments Dialog Box

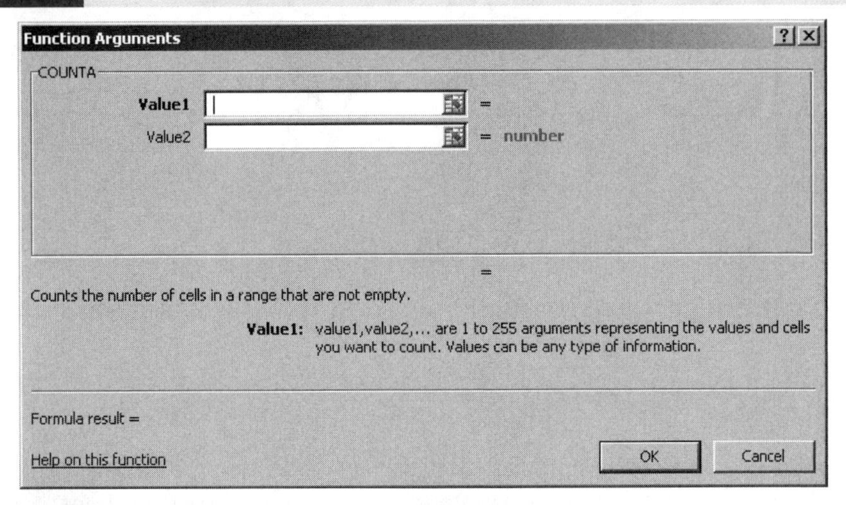

Figure 33.2 The COUNTA Function Returning the Number of Participants and Their Preferences

B28 =COUNTA(B2:B26)

	A	B	C	D	E
1	Respondent	Preference			
2	1	beans			
3	2	mac 'n' cheese			
4	3	mac 'n' cheese			
5	4				
6	5				
7	6	mac 'n' cheese			
8	7				
9	8	beans			
10	9				
11	10	mac 'n' cheese			
12	11				
13	12	mac 'n' cheese			
14	13				
15	14				
16	15	mac 'n' cheese			
17	16				
18	17	beans			
19	18				
20	19	mac 'n' cheese			
21	20				
22	21	beans			
23	22				
24	23	mac 'n' cheese			
25	24	mac 'n' cheese			
26	25	beans			
27					
28	COUNTA	14			

Check Your Understanding

To check your understanding of the COUNTA function, do the following two problems and check your answers in Appendix A.

QS 33a. Compute the COUNTA value for the number of people who decided to enroll in an elective class. Use Data Set 33a.

QS 33b. Compute the COUNTA value for the number of communities that adopted curbside or alley recycling. Use Data Set 33b.

EXCEL QUICKGUIDE 34

The COUNTBLANK Function

What the COUNTBLANK Function Does

The COUNTBLANK function returns the number of empty cells in a range of cells.

The Data Set

The data set is titled COUNTBLANK, and the question being asked is, "How many people who were contacted did not agree to participate in the experiment?"

Variable	*Value*
Agreed	1 (agreed) or empty cell

Using the COUNTBLANK Function

1. Click on the cell where you want the COUNTBLANK function to be placed. (In the data set, the cell is B23.)
2. Click the Formulas tab → Insert Function button (*fx*) and locate and double-click on the COUNTBLANK function. You will see the Function Arguments dialog box, as shown in Figure 34.1.
3. Click the RefEdit button in the Range: entry box and enter the range of the cells that contains the values you want included. In this example, the range is B2:B21.
4. Click the OK button. The COUNTBLANK function returns a value of 8 in Cell B23, as you see in Figure 34.2, indicating that 8 respondents did not agree to participate. Note that you can see the syntax for the function in the formula bar at the top of the worksheet.

Related Functions: AVERAGEIF, COUNT, COUNTA, COUNTIF

Figure 34.1 The COUNTBLANK Function Arguments Dialog Box

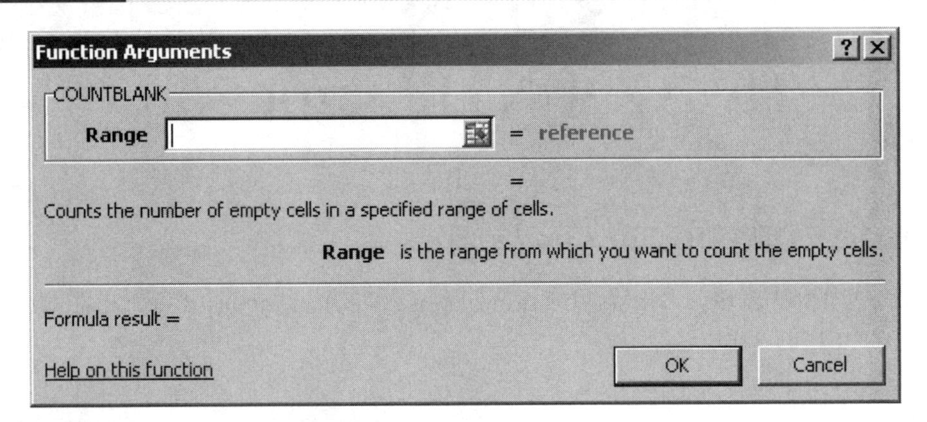

Figure 34.2 The COUNTBLANK Function Returning the Number of Respondents Who Did Not Agree to Participate

B23 | =COUNTBLANK(B2:B21)

	A	B	C	D	E
1	ID	Agreed			
2	1	1			
3	2				
4	3	1			
5	4	1			
6	5				
7	6				
8	7	1			
9	8				
10	9	1			
11	10	1			
12	11				
13	12				
14	13	1			
15	14	1			
16	15	1			
17	16	1			
18	17				
19	18	1			
20	19	1			
21	20				
22					
23	COUNTBLANK	8			

Check Your Understanding

To check your understanding of the COUNTBLANK function, do the following two problems and check your answers in Appendix A.

QS 34a. Compute the COUNTBLANK value for the number of students from a group of 35 who did not show up in an elective class. Use Data Set 34a.

QS 34b. Compute the COUNTBLANK value for the number of communities that do not have an adequate tax base. Use Data Set 34b.

EXCEL QUICKGUIDE 35

The COUNTIF Function

What the COUNTIF Function Does

The COUNTIF function counts the number of cells in a range of cells that meet a certain criterion.

The Data Set

The data set is titled COUNTIF, and the question being asked is, "How many business transactions were evaluated as successful?"

Variable	*Value*
Successful	1 = Yes, 2 = No

Using the COUNTIF Function

1. Click on the cell where you want the COUNTIF function to be placed. (In the data set, the cell is B23.)
2. Click the Formulas tab → Insert Function button (*fx*) and locate and double-click on the COUNTIF function. You will see the Function Arguments dialog box, as shown in Figure 35.1.
3. Click the RefEdit button in the Range entry box and enter the range of the cells that contains the values you want included. In this example, the range is B2:B21.
4. Click the OK button. The COUNTIF function returns a value of 13 in Cell B23, as you see in Figure 35.2, indicating that 13 transactions were successful. Note that you can see the syntax for the function in the formula bar at the top of the worksheet.

Related Functions: AVERAGEIF, COUNT, COUNTA, COUNTBLANK

Figure 35.1 The COUNTIF Function Arguments Dialog Box

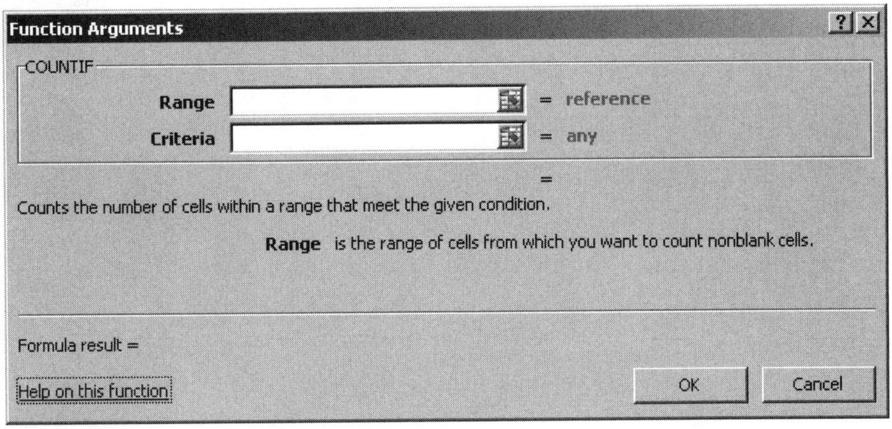

Function Arguments

COUNTIF

Range = reference

Criteria = any

=

Counts the number of cells within a range that meet the given condition.

Range is the range of cells from which you want to count nonblank cells.

Formula result =

Help on this function

OK Cancel

Figure 35.2 The COUNTIF Function Returning the Number of Successful Transactions

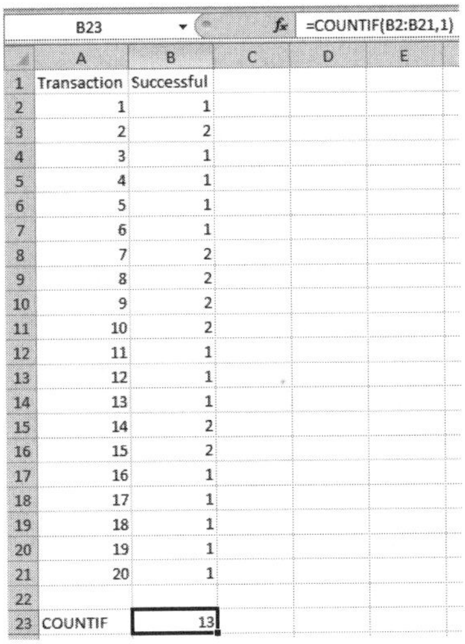

B23 =COUNTIF(B2:B21,1)

	A	B	C	D	E
1	Transaction	Successful			
2	1	1			
3	2	2			
4	3	1			
5	4	1			
6	5	1			
7	6	1			
8	7	2			
9	8	2			
10	9	2			
11	10	2			
12	11	1			
13	12	1			
14	13	1			
15	14	2			
16	15	2			
17	16	1			
18	17	1			
19	18	1			
20	19	1			
21	20	1			
22					
23	COUNTIF	13			

Check Your Understanding

To check your understanding of the COUNTIF function, do the following two problems and check your answers in Appendix A.

QS 35a. Compute the COUNTIF value for the number of students who were admitted to the program and subsequently enrolled (1 = Did Enroll, 2 = Did Not Enroll). Use Data Set 35a.

QS 35b. Compute the COUNTIF value for the number of businesses that are open on the weekends (1 = Yes, 2 = No). Use Data Set 35b.

PART II

USING THE ANALYSIS TOOLPAK

The Excel Analysis ToolPak is a set of automated tools that allows you to conduct a variety of simple and complex statistical analyses. They are similar to Excel functions in that each requires input from the user, but they produce much more elaborate output rather than just one outcome (as do most functions).

For example, the Descriptive Statistics tool in the Analysis ToolPak produces the mean, median, and mode, among other descriptive measures. To compute such values using functions, you would have to use at least three different functions. With the ToolPak, you can do it (and more) by using only one.

The basic method for using any one of the ToolPak tools is as follows:

1. From the Data tab, select the Data Analysis option.
2. Select the tool you want to use.
3. Enter the data and the type of the analysis you want done.
4. Format the data as you see fit. In the examples used in this part of *Excel Statistics: A Quick Guide,* the data may have been reformatted to appear more pleasing to the eye and to be more internally consistent (for example, numbers are formatted to two decimal places, labels may be centered when appropriate, etc.).

The process is very much like using a function, including use of the RefEdit button to enter data into the dialog box, but you may have to make a number of decisions as you prepare the analysis. Among these are the following:

1. Whether you want labels to be used in the ToolPak output
2. Whether you want the output to appear on the current Excel worksheet, on a new worksheet, or in an entirely new workbook
3. Whether the results should be grouped by columns or rows
4. The range of cells in which you want the output to appear
5. Whether you will enter data in the Data Analysis dialog box rather than using the RefEdit button and dragging over the data

For our purposes here, we are always going to use labels, will have the output appear on the same worksheet, and will group the output by columns. The range of cells in which the output should appear will be defined as each Analysis ToolPak tool is discussed. The final or partial output you see will always be reformatted to better fit the page.

Once the Analysis ToolPak output is complete, it can be manipulated as any Excel data can. In addition to being cut and paste into other applications, such as Word, the output can easily be modified using Excel's Format as Table option on the Home tab. For example, the simple output you saw in Figure 17.2 can easily be reformatted to appear as shown in Figure II.1.

Figure II.1 Modifying Excel Analysis ToolPak Output

ID	Height	Weight
1	60	134
2	63	143
3	71	156
4	58	121
5	61	131
6	59	117
7	64	125
8	67	126
9	63	143
10	52	98
11	61	154
12	58	125
13	54	109
14	61	117
15	64	126
16	63	154
17	49	98
18	59	143
19	69	144
20	71	156
CORREL		0.776

EXCEL QUICKGUIDE 36

Descriptive Statistics

What the Descriptive Statistics Tool Does

The Descriptive Statistics tool computes basic descriptive statistics for a set of data, such as the mean, median, and mode among others.

The Data Set

The data set used in this example is titled DESCRIPTIVE STATISTICS, and the question is, "What are the descriptive statistics for height for a group of 20 two-year-olds?"

Variable	*Description*
Height	Height in inches of 2-year-olds

Using the Descriptive Statistics Tool

1. Select the Data Analysis option from the Data tab.
2. Double-click on the Descriptive Statistics option in the Data Analysis dialog box, and you will see the Descriptive Statistics dialog box, as shown in Figure 36.1.
3. Define the Input Range by clicking on the RefEdit button and selecting the data you want to use in the analysis (in this example, Cells A1 through A21). Click the RefEdit button again. Be sure to check the Labels in First Row box.
4. Click the Output Range button, click the RefEdit button, and define the output range by selecting an area in the worksheet where you want the output to appear (in this example, Cell C2). Click the RefEdit button once again. Even though the range is only one cell, Excel will know to extend it to fit all the output.
5. Click the Summary statistics option and click OK.

The Final Output

The Descriptive Statistics output, including the original data and the summary statistics, is shown in Figure 36.2. Note that the cells were formatted where appropriate using the Format Cells command. Otherwise, Excel produced what you see as the final output.

Figure 36.1 The Descriptive Statistics Dialog Box

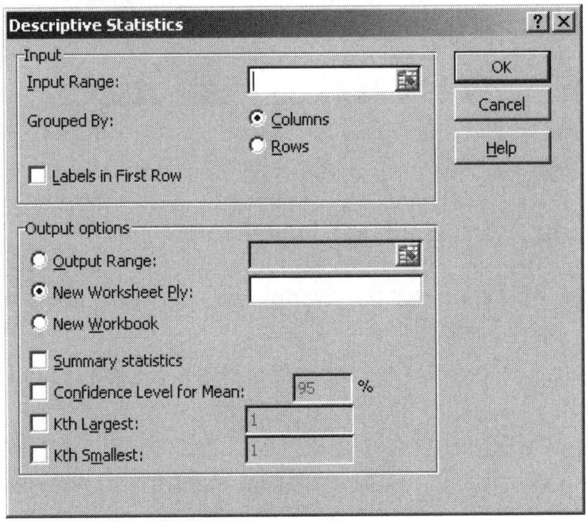

Descriptive Statistics
Input
Input Range:
Grouped By: Columns / Rows
Labels in First Row
Output options
Output Range:
New Worksheet Ply:
New Workbook
Summary statistics
Confidence Level for Mean: 95 %
Kth Largest: 1
Kth Smallest: 1
OK
Cancel
Help

Figure 36.2 The Descriptive Statistics Output

	A	B	C	D
1	Height			
2	38		*Height*	
3	28			
4	28		Mean	32.7
5	40		Standard Error	1.04
6	29		Median	33.50
7	36		Mode	28.00
8	35		Standard Deviation	4.64
9	28		Sample Variance	21.48
10	38		Kurtosis	-1.35
11	37		Skewness	0.08
12	35		Range	14
13	34		Minimum	26
14	26		Maximum	40
15	26		Sum	654
16	40		Count	20
17	31			
18	30			
19	34			
20	33			
21	28			

Check Your Understanding

To check your understanding of the Descriptive Statistics tool, do the following two problems and check your answers in Figures A.1 and A.2 in Appendix A.

QS 36a. Compute the descriptive statistics for a group of cities and their sales tax rates. Use Data Set 36a.

QS 36b. Compute the descriptive statistics for number of car sales for 4 weeks in the month of June. Use Data Set 36b.

EXCEL QUICKGUIDE 37

Moving Average

What the Moving Average Tool Does

The Moving Average tool computes the average for sets of numbers at a defined interval.

The Data Set

The data set used in this example is titled MOVING AVERAGE, and the question is, "What is the moving average of number of home sales for subsequent 4-week periods?"

Variable	*Description*
Sales	Number of home sales in a 1-week period

Using the Moving Average Tool

1. Select the Data Analysis option from the Data tab.
2. Double-click on the Moving Average option in the Data Analysis dialog box, and you will see the Moving Average dialog box shown in Figure 37.1.
3. Define the Input Range by clicking on the RefEdit button and selecting the data you want to use in the analysis (in this example, B1 through B13). Click the RefEdit button once again. Be sure to check the Labels in First Row box.
4. Define the Interval, or the number of values you want to be included in the calculation of each average. In this example, enter 4 directly in the dialog box.
5. Click the Output Range button, click the RefEdit button, and select an area in the worksheet where you want the output to appear (in this example, D1). Click the RefEdit button once again. Even though the range is only one cell, Excel will know to extend it to fit all the output.
6. Check the Chart Output checkbox and click OK.

The Final Output

The Moving Average output, including the original data and the averages, is shown in Figure 37.2. As you can see, the first three averages are marked #N/A because Excel must detect four averages before it can compute the first one. Otherwise, Excel produces what you see as the final output.

Figure 37.1 The Moving Average Dialog Box

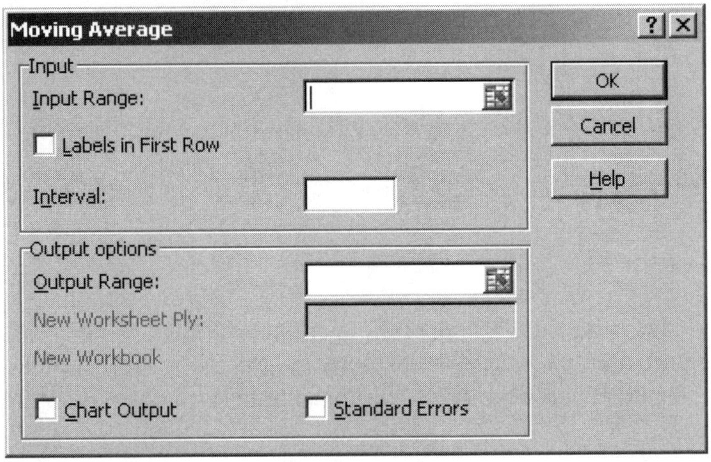

Moving Average

Input
Input Range:
Labels in First Row
Interval:
Output options
Output Range:
New Worksheet Ply:
New Workbook
Chart Output
Standard Errors
OK
Cancel
Help

Figure 37.2 The Moving Average Output

	A	B	C	D
1	Week	Sales		#N/A
2	1	$4,567.00		#N/A
3	2	$3,454.00		#N/A
4	3	$5,456.00		$ 5,278.00
5	4	$7,635.00		$ 5,239.50
6	5	$4,413.00		$ 5,540.25
7	6	$4,657.00		$ 6,242.25
8	7	$8,264.00		$ 5,177.50
9	8	$3,376.00		$ 4,938.25
10	9	$3,456.00		$ 4,915.25
11	10	$4,565.00		$ 3,407.75
12	11	$2,234.00		$ 4,200.25
13	12	$6,546.00		

Check Your Understanding

To check your understanding of the MOVING AVERAGE tool, do the following two problems and check your answers in Figures A.3 and A.4 in Appendix A.

QS 37a. Compute the Moving Average of use of parking spaces in a 100-space lot for adjacent 4-month periods. Use Data Set 37a.

QS 37b. Compute the Moving Average of hours training for each 5-week period over 12 weeks. Use Data Set 37b.

EXCEL QUICKGUIDE 38

Random Number Generation

What the Random Number Generation Tool Does

The Random Number Generation tool produces a set of random numbers.

The Data Set

The data set used in this example is titled RANDOM NUMBER GENERATOR, and the tool generates a random number for each of 20 members of a group. The numbers will be used to assign members to experimental (odd number) or control (even number) groups.

Variable	*Description*
ID	Participant's identifying number
Random Number	Random number assigned to each ID

Using the Random Number Generation Tool

1. Select the Data Analysis option from the Data tab.
2. Double-click on the Random Number Generation option in the Data Analysis dialog box, and you will see the Random Number Generation dialog box shown in Figure 38.1.
3. Define the Number of Variables (in this example, 1).
4. Define the Number of Random Numbers (in this example, 20).
5. Select Normal from the Distribution drop-down menu. The Random Number Generation dialog box changes, offering you entry boxes for Mean and Standard deviation.
6. Enter 1 for Mean.
7. Enter 1 for Standard deviation.
8. Click on the RefEdit button for the Output Range and select the range where you want the random numbers to be placed (in this example, it is B2 through B21). Click the RefEdit button.
9. Click OK.

The Final Output

The 20 random digits appear as shown in Figure 38.2. Members with numbers ending in an odd digit are to be placed in the experimental group, and members with numbers ending in an even digit will be placed in the control group.

Figure 38.1 The Random Number Generation Dialog Box

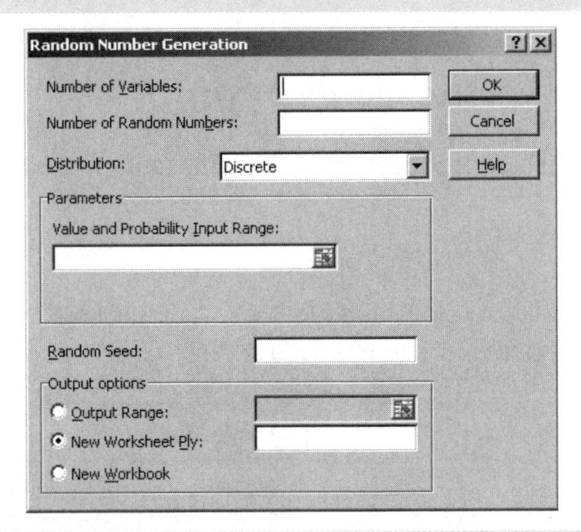

Figure 38.2 The Random Number Generation Output

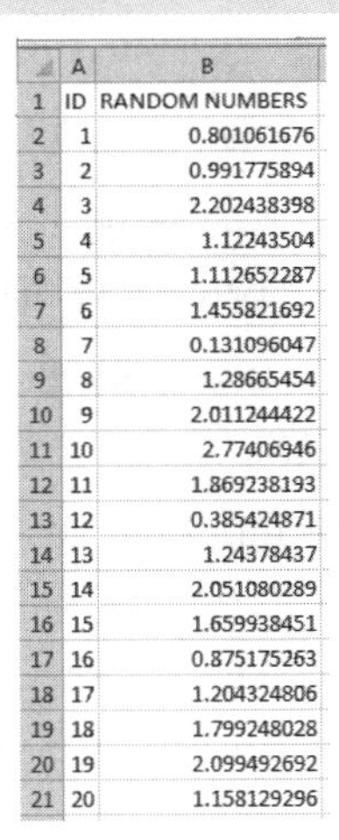

	A	B
1	ID	RANDOM NUMBERS
2	1	0.801061676
3	2	0.991775894
4	3	2.202438398
5	4	1.12243504
6	5	1.112652287
7	6	1.455821692
8	7	0.131096047
9	8	1.28665454
10	9	2.011244422
11	10	2.77406946
12	11	1.869238193
13	12	0.385424871
14	13	1.24378437
15	14	2.051080289
16	15	1.659938451
17	16	0.875175263
18	17	1.204324806
19	18	1.799248028
20	19	2.099492692
21	20	1.158129296

Check Your Understanding

To check your understanding of the Random Number Generation tool, do the following two problems and check your answers in Figures A.5 and A.6 in Appendix A.

QS 38a. Compute a set of 25 random numbers. Use Data Set 38a.

QS 38b. Compute a set of 10 random numbers to assign each of 10 participants to one of two groups in an experiment. Use Data Set 38b.

EXCEL QUICKGUIDE 39

Rank and Percentile

What the Rank and Percentile Tool Does

The Rank and Percentile tool produces the point, rank, and percentiles for a set of scores.

The Data Set

The data set used in this example is titled RANK AND PERCENTILE, and the question is, "What are the rank and percentile scores for a set of 10 grade point averages (GPAs)?"

Variable	*Description*
GPA	Grade point average ranging from 0.0 to 4.0

Using the Rank and Percentile Tool

1. Select the Data Analysis option from the Data tab.
2. Double-click on the Rank and Percentile option in the Data Analysis dialog box, and you will see the Rank and Percentile dialog box shown in Figure 39.1.
3. Be sure that the Columns button and the Labels in First Row checkbox are selected.
4. Click the RefEdit button in the Input Range and select the data you want to use in the analysis (in this example, it is A1 through A11). Click the RefEdit button again.
5. Click the RefEdit button in the Output Range and select the location where you want the output to appear (in this example, B1). Click the RefEdit button again.
6. Click OK.

The Final Output

The point (the GPA's location in the original set of data), the GPA, the rank, and the percent (percentile) for each GPA in the data set are shown in Figure 39.2.

Figure 39.1 The Rank and Percentile Dialog Box

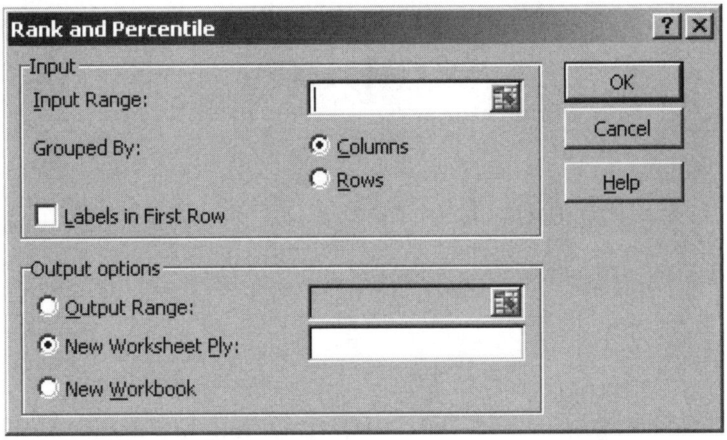

Rank and Percentile
Input
Input Range:
Grouped By: Columns
Rows
Labels in First Row
Output options
Output Range:
New Worksheet Ply:
New Workbook
OK
Cancel
Help

Figure 39.2 The Rank and Percentile Output

	A	B	C	D	E
1	GPA	*Point*	*GPA*	*Rank*	*Percent*
2	3.8	6	3.9	1	100.00%
3	2.5	1	3.8	2	88.80%
4	3.1	9	3.5	3	77.70%
5	1.6	3	3.1	4	66.60%
6	2.8	5	2.8	5	55.50%
7	3.9	7	2.6	6	44.40%
8	2.6	2	2.5	7	33.30%
9	1.2	10	2.2	8	22.20%
10	3.5	4	1.6	9	11.10%
11	2.2	8	1.2	10	0.00%

Check Your Understanding

To check your understanding of the Rank and Percentile tool, do the following two problems and check your answers in Figures A.7 and A.8 in Appendix A.

QS 39a. Compute the rank and percentile for a group of 20 salaries. Use Data Set 39a.

QS 39b. Compute the rank and percentile for 15 students and the hours studied per week. Use Data Set 39b.

EXCEL QUICKGUIDE 40

Sampling

What the Sampling Tool Does

The Sampling tool produces a sample that is selected from a population.

The Data Set

The data set used in this example is titled SAMPLING, and the question is, "What are the identification numbers of 10 participants selected at random from a population of 100?"

Variable	*Description*
ID	Identification number

Using the Sampling Tool

1. Select the Data Analysis option from the Data tab.
2. Double-click on the Sampling option in the Data Analysis dialog box, and you will see the Sampling dialog box shown in Figure 40.1.
3. Click the RefEdit button in the Input Range entry box and select the data you want to use in the analysis (in this example, it is A2 through E21). Click the RefEdit button again.
4. Click the Random button and enter the number you want randomly selected. In this example, the number is 10.
5. Click the RefEdit button in the Output Range and select the location where you want the output to appear (in this example, G2). Click the RefEdit button again.
6. Click OK.

The Final Output

The 10 ID numbers randomly selected are shown in Figure 40.2.

Figure 40.1 The Sampling Dialog Box

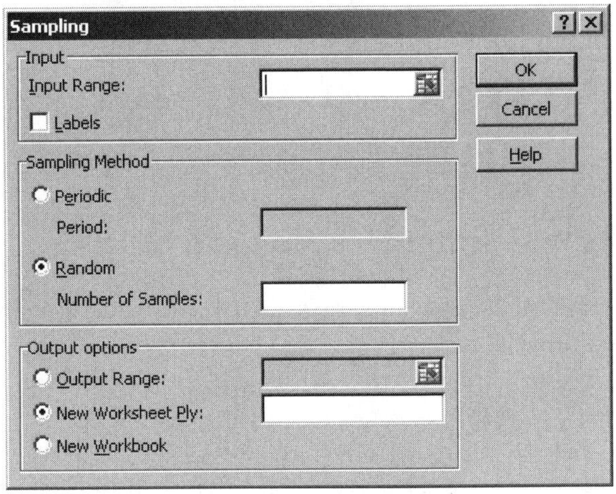

Sampling

Input
Input Range:
Labels

Sampling Method
Periodic
Period:
Random
Number of Samples:

Output options
Output Range:
New Worksheet Ply:
New Workbook

OK
Cancel
Help

Figure 40.2 The Sampling Output

	A	B	C	D	E	F	G
1	ID						Sample
2	11	70	68	18	21		41
3	97	83	13	69	84		3
4	35	98	14	77	18		32
5	26	76	13	67	41		21
6	79	5	93	20	91		70
7	48	76	42	19	98		76
8	37	28	56	90	18		67
9	86	72	12	41	34		7
10	30	80	28	58	64		98
11	14	50	14	13	9		28
12	15	73	58	55	69		
13	9	55	48	44	36		
14	7	11	3	22	70		
15	32	69	3	68	23		
16	89	9	18	96	76		
17	92	23	35	90	59		
18	44	65	95	76	26		
19	86	66	65	69	10		
20	70	55	100	37	23		
21	42	16	30	89	57		

Check Your Understanding

To check your understanding of the Sampling tool, do the following two problems and check your answers in Figures A.9 and A.10 in Appendix A.

QS 40a. Select a random sample of 15 participants from a population of 90. Use Data Set 40a.

QS 40b. Select a random sample of 20 from a population of 200. Use Data Set 40b.

EXCEL QUICKGUIDE 41

z-Test: Two-Sample for Means

What the z-Test: Two-Sample for Means Tool Does

The z-Test: Two-Sample for Means computes a z value between means when the population variances are known.

The Data Set

The data set used in this example is titled ZTEST, and the question is, "Do the population means for urban and rural residents differ on a test of energy use?"

Variable	*Description*
Energy Use	Total annual energy costs in dollars

Using the z-Test: Two-Sample for Means Tool

1. Select the Data Analysis option from the Data tab.
2. Double-click on the z-Test: Two-Sample for Means option in the Data Analysis dialog box, and you will see the z-Test: Two-Sample for Means dialog box shown in Figure 41.1.
3. Be sure that the Labels box in the Input area is checked.
4. Click the RefEdit button in the Variable 1 Range entry box and select the data you want to use in the analysis. In this example, it is Cells B1 through B21. Click the RefEdit button again.
5. Repeat Step 4 for the Variable 2 Range entry box and select the data you want to use in the analysis. In this example, it is Cells C1 through C21. Click the RefEdit button again.
6. Enter the known population variance for both arrays of data in Variable 1 Variance (known) and Variable 2 Variance (known), respectively. In this example, VAR.P was used to compute these values, which are available in Cells B23 and C23.
7. Click the RefEdit button and enter the output range. In this example, Cell D1 was selected. Click the RefEdit button again.
8. Click OK.

The Final Output

The result, shown in Cell E8 of Figure 41.2, is a z value of 5.12, which is significant beyond the .05 level. This indicates that urban and rural residents differ in their energy use.

Figure 41.1 The z-Test: Two-Sample for Means Dialog Box

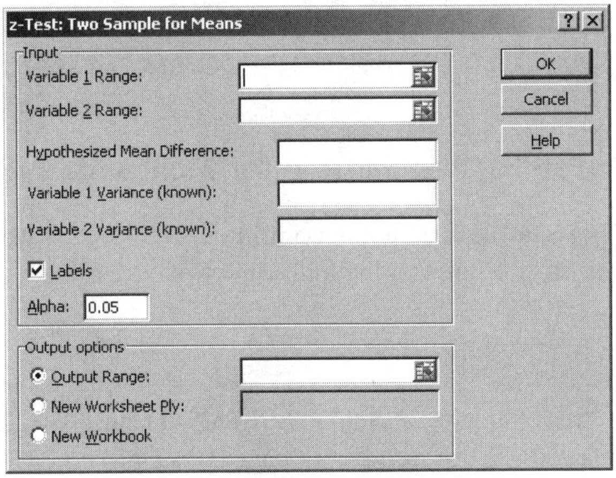

z-Test: Two Sample for Means

Input
Variable 1 Range:
Variable 2 Range:
Hypothesized Mean Difference:
Variable 1 Variance (known):
Variable 2 Variance (known):
☑ Labels
Alpha: 0.05

Output options
◉ Output Range:
○ New Worksheet Ply:
○ New Workbook

OK
Cancel
Help

Figure 41.2 The z-Test: Two-Sample for Means Output

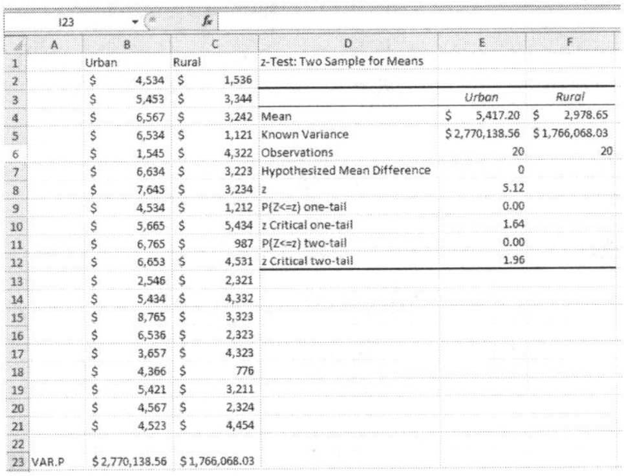

	A	B	C	D	E	F
1		Urban	Rural	z-Test: Two Sample for Means		
2		$ 4,534	$ 1,536			
3		$ 5,453	$ 3,344		*Urban*	*Rural*
4		$ 6,567	$ 3,242	Mean	$ 5,417.20	$ 2,978.65
5		$ 6,534	$ 1,121	Known Variance	$ 2,770,138.56	$ 1,766,068.03
6		$ 1,545	$ 4,322	Observations	20	20
7		$ 6,634	$ 3,223	Hypothesized Mean Difference	0	
8		$ 7,645	$ 3,234	z	5.12	
9		$ 4,534	$ 1,212	P(Z<=z) one-tail	0.00	
10		$ 5,665	$ 5,434	z Critical one-tail	1.64	
11		$ 6,765	$ 987	P(Z<=z) two-tail	0.00	
12		$ 6,653	$ 4,531	z Critical two-tail	1.96	
13		$ 2,546	$ 2,321			
14		$ 5,434	$ 4,332			
15		$ 8,765	$ 3,323			
16		$ 6,536	$ 2,323			
17		$ 3,657	$ 4,323			
18		$ 4,366	$ 776			
19		$ 5,421	$ 3,211			
20		$ 4,567	$ 2,324			
21		$ 4,523	$ 4,454			
22						
23	VAR.P	$ 2,770,138.56	$ 1,766,068.03			

Check Your Understanding

To check your understanding of the z-Test: Two-Sample for Means tool, do the following two problems and check your answers in Figures A.11 and A.12 in Appendix A.

QS 41a. Compute the z value and evaluate it for significance for the scores on an attitude scale ranging from 1 to 5 (with 5 being most satisfied) between two groups of 20 homeowners. Use Data Set 41a.

QS 41b. Compute the z value and evaluate it for significance for the heaviest single lift between two groups of 15 participants in a program to increase strength. Use Data Set 41b.

EXCEL QUICKGUIDE 42

t-Test: Paired Two-Sample for Means

What the t-Test: Paired Two-Sample for Means Tool Does

The t-Test: Paired Two-Sample for Means computes a *t* value between means for two dependent measures on the same individuals or set of cases.

The Data Set

The data set used in this example is titled TTEST-PAIRED, and the question is "Does an intervention program reduce the number of cigarettes smoked each day?"

Variable	*Description*
Before	Number of cigarettes smoked before the intervention
After	Number of cigarettes smoked after the intervention

Using the t-Test: Paired Two-Sample for Means Tool

1. Select the Data Analysis option from the Data tab.
2. Double-click on the t-Test: Paired Two-Sample for Means option in the Data Analysis dialog box, and you will see the t-Test: Paired Two-Sample for Means dialog box shown in Figure 42.1.
3. Click the RefEdit button in the Variable 1 Range entry box and select the data you want to use in the analysis. In this example, it is Cells A1 through A21. Click the RefEdit button again.
4. Repeat Step 3 for the Variable 2 Range entry box and select the data you want to use in the analysis. In this example, it is Cells B1 through B21. Click the RefEdit button again.
5. Be sure that the Labels box in the Input area is checked.
6. Be sure that an Alpha value of 0.05 is selected.
7. Click the RefEdit button and enter the Output Range. In this example, Cell C1 is selected. Click the RefEdit button again.
8. Click OK.

The Final Output

The result, shown in Figure 42.2, is a t value of 6.71, which is significant beyond the .05 level (Alpha value in Figure 42.1). This indicates that the intervention was effective and there was a difference in rate of daily cigarette smoking.

Figure 42.1 The t-Test: Paired Two-Sample for Means Dialog Box

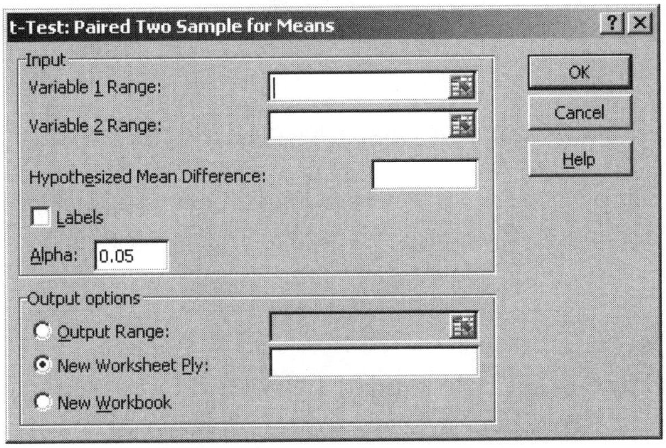

t-Test: Paired Two Sample for Means

Input
Variable 1 Range:
Variable 2 Range:
Hypothesized Mean Difference:
Labels
Alpha: 0.05

Output options
Output Range:
New Worksheet Ply:
New Workbook

OK
Cancel
Help

Figure 42.2 The t-Test: Paired Two-Sample for Means Output

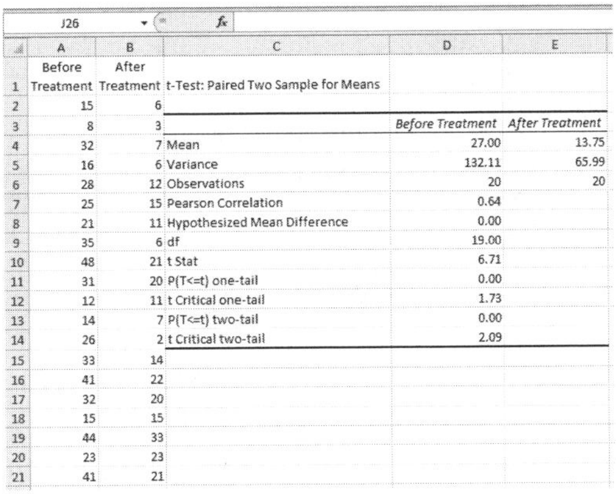

J26

	A	B	C	D	E
1	Before Treatment	After Treatment	t-Test: Paired Two Sample for Means		
2	15	6			
3	8	3		*Before Treatment*	*After Treatment*
4	32	7	Mean	27.00	13.75
5	16	6	Variance	132.11	65.99
6	28	12	Observations	20	20
7	25	15	Pearson Correlation	0.64	
8	21	11	Hypothesized Mean Difference	0.00	
9	35	6	df	19.00	
10	48	21	t Stat	6.71	
11	31	20	P(T<=t) one-tail	0.00	
12	12	11	t Critical one-tail	1.73	
13	14	7	P(T<=t) two-tail	0.00	
14	26	2	t Critical two-tail	2.09	
15	33	14			
16	41	22			
17	32	20			
18	15	15			
19	44	33			
20	23	23			
21	41	21			

Check Your Understanding

To check your understanding of the t-Test: Paired Two-Sample for Means tool, do the following two problems and check your answers in Figures A.13 and A.14 in Appendix A.

QS 42a. Compute the t value and evaluate it for significance for a law school admission scores for the same group of 25 participants before and after a crash study course. Use Data Set 42a.

QS 42b. Compute the t value and evaluate it for significance for body mass index scores (BMI) for 20 participants in a wellness program before and after the program. Use Data Set 42b.

EXCEL QUICKGUIDE 43

t-Test: Two-Sample Assuming Unequal Variances

What the t-Test: Two-Sample Assuming Unequal Variances Tool Does

The t-Test: Two-Sample Assuming Unequal Variances computes a *t* value between means for two independent measures when the variances for each group are unequal.

The Data Set

The data set used in this example is titled TTEST-UNEQUAL, and the question is, "Is there a difference in contribution levels to nonprofits between married and never-married females?"

Variable	*Description*
Married Females	Amount of money donated to not-for-profit organizations in dollars per year by married females
Never Married Females	Amount of money donated to not-for-profit organizations in dollars per year by never-married females

Using the t-Test: Two-Sample Assuming Unequal Variances Tool

1. Select the Data Analysis option from the Data tab.
2. Double-click on the t-Test: Two-Sample Assuming Unequal Variances option in the Data Analysis dialog box, and you will see the t-Test: Two-Sample Assuming Unequal Variances dialog box shown in Figure 43.1.
3. Click the RefEdit button in the Variable 1 Range entry box and select the data you want to use in the analysis. In this example, it is Cells A1 through A21. Click the RefEdit button again.
4. Repeat Step 3 for the Variable 2 Range entry box and select the data you want to use in the analysis. In this example, it is Cells B1 through B23. Click the RefEdit button again.
5. Be sure that the Labels box in the Input area is checked.

6. Click the RefEdit button and enter the Output Range. In this example, Cell C1 was selected. Click the RefEdit button again.
7. Click OK.

The Final Output

The result, shown in Figure 43.2, is a *t* value of −0.29, which is not significant beyond the .05 level. This indicates that there is no difference in the amount of dollars contributed to nonprofit organizations by married and never-married females.

Figure 43.1 The t-Test: Two-Sample Assuming Unequal Variances Dialog Box

t-Test: Two-Sample Assuming Unequal Variances
Input
Variable 1 Range:
Variable 2 Range:
Hypothesized Mean Difference:
Labels
Alpha: 0.05
Output options
Output Range:
New Worksheet Ply:
New Workbook
OK
Cancel
Help

Figure 43.2 The t-Test: Two-Sample Assuming Unequal Variances Output

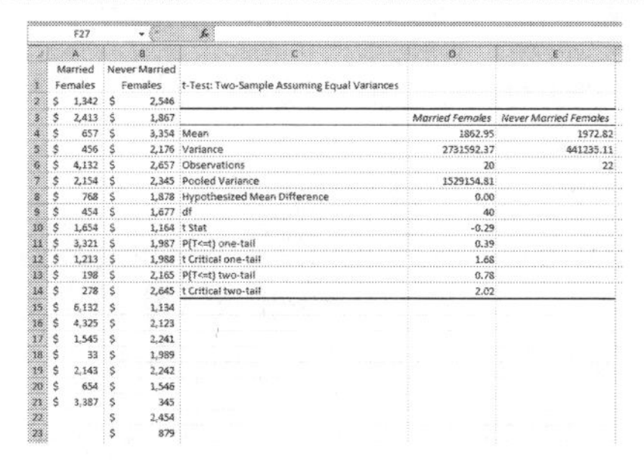

	A	B	C	D	E
1	Married Females	Never Married Females	t-Test: Two-Sample Assuming Equal Variances		
2	$ 1,342	$ 2,546			
3	$ 2,413	$ 1,867		*Married Females*	*Never Married Females*
4	$ 657	$ 3,354	Mean	1862.95	1972.82
5	$ 456	$ 2,176	Variance	2731592.37	441235.11
6	$ 4,132	$ 2,657	Observations	20	22
7	$ 2,154	$ 2,345	Pooled Variance	1529154.81	
8	$ 768	$ 1,878	Hypothesized Mean Difference	0.00	
9	$ 454	$ 1,677	df	40	
10	$ 1,654	$ 1,164	t Stat	-0.29	
11	$ 3,321	$ 1,987	P(T<=t) one-tail	0.39	
12	$ 1,213	$ 1,988	t Critical one-tail	1.68	
13	$ 198	$ 2,165	P(T<=t) two-tail	0.78	
14	$ 278	$ 2,645	t Critical two-tail	2.02	
15	$ 6,132	$ 1,134			
16	$ 4,325	$ 2,123			
17	$ 1,545	$ 2,241			
18	$ 33	$ 1,989			
19	$ 2,143	$ 2,242			
20	$ 654	$ 1,546			
21	$ 3,387	$ 345			
22		$ 2,454			
23		$ 879			

Check Your Understanding

To check your understanding of the t-Test: Two-Sample Assuming Unequal Variables, do the following two problems and check your answers in Figures A.15 and A.16 in Appendix A.

QS 43a. Compute the *t* value and evaluate it for the significance in differences in infant mortality rates (per 1,000 live births) between two different groups of countries. Use Data Set 43a.

QS 43b. Compute the *t* value and evaluate it for the significance in differences in amount of student debt after 10 years between private and public debt. Use Data Set 43b.

EXCEL QUICKGUIDE 44

t-Test: Two-Sample Assuming Equal Variances

What the t-Test: Two-Sample Assuming Equal Variances Tool Does

The t-Test: Two-Sample Assuming Equal Variances computes a *t* value between means for two independent measures when the variances for each group are equal.

The Data Set

The data set used in this example is titled TTEST-EQUAL, and the question is, "Is there a difference in weekly sales of units between Region 1 and Region 2?"

Variable	*Description*
Sales R1	Weekly sales in units for Region 1
Sales R2	Weekly sales in units for Region 2

Using the t-Test: Two-Sample Assuming Equal Variances Tool

1. Select the Data Analysis option from the Data tab.
2. Double-click on the t-Test: Two-Sample Assuming Equal Variances option in the Data Analysis dialog box, and you will see the t-Test: Two-Sample Assuming Equal Variances dialog box shown in Figure 44.1.
3. Click the RefEdit button in the Variable 1 Range entry box and select the data you want to use in the analysis. In this example, it is Cells B1 through B17. Click the RefEdit button.
4. Repeat Step 3 for the Variable 2 Range entry box and select the data you want to use in the analysis. In this example, it is Cells C1 through C17. Click the RefEdit button.
5. Be sure that the Labels box in the Input area is checked.
6. Click the RefEdit button and enter the Output Range. In this example, Cell D1 was selected. Click the RefEdit button again.
7. Click OK.

The Final Output

The result, shown in Figure 44.2, is a t value of −2.76, which is significant beyond the .05 level (as specified in the dialog box you see in Figure 44.1) for a two-tailed test. This indicates that there is a difference in sales between the two regions.

Figure 44.1 The t-Test: Two-Sample Assuming Equal Variances Dialog Box

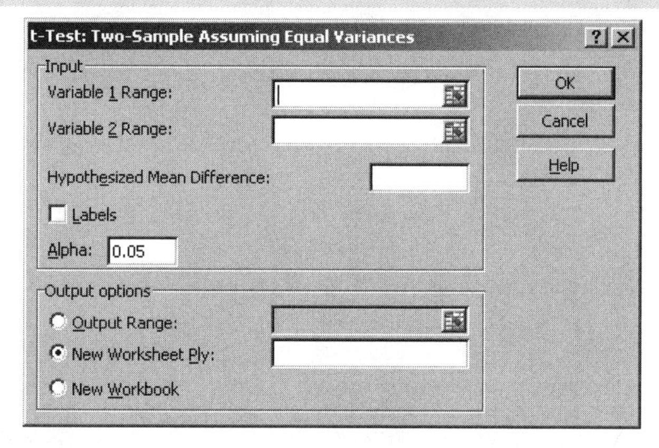

Figure 44.2 The t-Test: Two-Sample Assuming Equal Variances Output

Week	Sales R1	Sales R2	t-Test: Two-Sample Assuming Equal Variances		
1	$ 465	$ 657			
2	$ 465	$ 789		*Sales R1*	*Sales R2*
3	$ 423	$ 456	Mean	408.56	531.69
4	$ 398	$ 655	Variance	3013.06	28745.56
5	$ 387	$ 354	Observations	16	16
6	$ 416	$ 775	Pooled Variance	15879.31	
7	$ 401	$ 726	Hypothesized Mean Difference	0.00	
8	$ 326	$ 389	df	30	
9	$ 345	$ 456	t Stat	-2.76	
10	$ 476	$ 331	P(T<=t) one-tail	0.00	
11	$ 312	$ 412	t Critical one-tail	1.70	
12	$ 334	$ 588	P(T<=t) two-tail	0.01	
13	$ 431	$ 312	t Critical two-tail	2.04	
14	$ 478	$ 331			
15	$ 453	$ 589			
16	$ 427	$ 687			

Check Your Understanding

To check your understanding of the t-Test: Two Samples Assuming Equal Variances, do the following two problems and check your answers in Figures A.17 and A.18 in Appendix A.

QS 44a. Compute the t value and evaluate it for significance for the difference between the number of cigarettes smoked each day between two groups, one of which is in a smoking-cessation program and the other is a control group. Use Data Set 44a.

QS 44b. Compute the t value and evaluate it for significance for the difference between 25 salaries of graduates of online versus bricks-and-mortar master's degree programs. Use Data Set 44b.

EXCEL QUICKGUIDE 45

Anova: Single Factor

What the Anova: Single Factor Tool Does

The Anova: Single Factor tests for differences between the means of two or more groups.

The Data Set

The data set used in this example is titled Anova-Single Factor, and the question is, "Is there a difference in language proficiency (LP) as a function of number of hours of weekly practice?"

Variable	*Description*
LP	Language proficiency
Practice	Number of hours of practice as either 5 or 10 per week

Using the Anova: Single Factor Tool

1. Select the Data Analysis option from the Data tab.
2. Double-click on the Anova: Single Factor option in the Data Analysis dialog box, and you will see the Anova: Single Factor dialog box shown in Figure 45.1.
3. Click the RefEdit button in the Input Range entry box and select the data you want to use in the analysis. In this example, it is Cells A1 through C21. Click the RefEdit button again.
4. Be sure to check the Labels in First Row box.
5. Click the RefEdit button and enter the Output Range. In this example, Cell E1 was selected. Click the RefEdit button again.
6. Click OK.

The Final Output

The result, shown in Figure 45.2, shows language proficiency increasing with more practice and the difference between the means of the three groups being significant beyond the .02 level, given an *F* value of 4.45.

Figure 45.1 The Anova: Single Factor Dialog Box

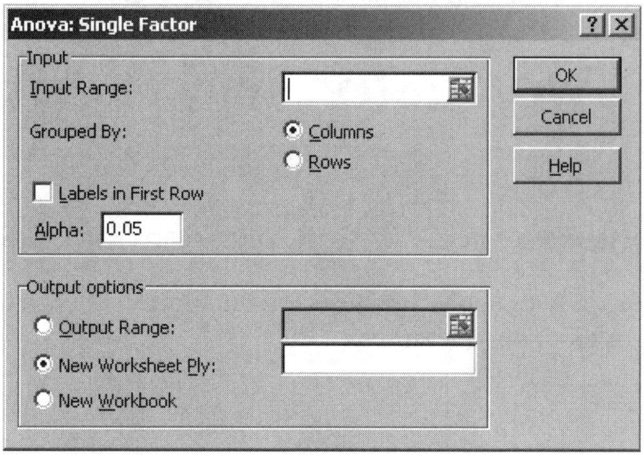

Anova: Single Factor
Input
Input Range:
Grouped By: Columns / Rows
Labels in First Row
Alpha: 0.05
Output options
Output Range:
New Worksheet Ply:
New Workbook
OK
Cancel
Help

Figure 45.2 The Anova: Single Factor Output

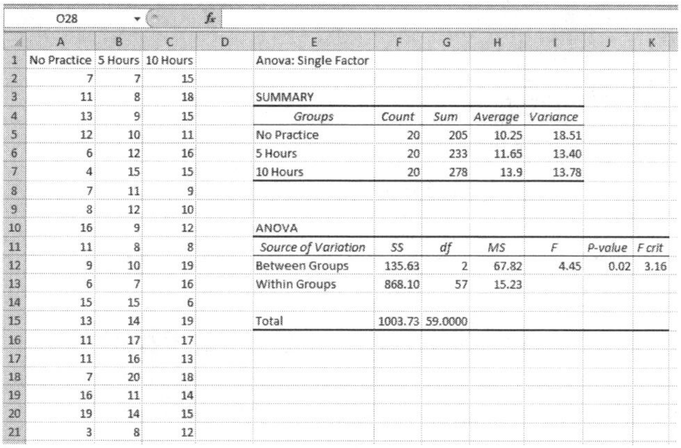

O28

	A	B	C	D	E	F	G	H	I	J	K
1	No Practice	5 Hours	10 Hours		Anova: Single Factor						
2	7	7	15								
3	11	8	18		SUMMARY						
4	13	9	15		*Groups*	*Count*	*Sum*	*Average*	*Variance*		
5	12	10	11		No Practice	20	205	10.25	18.51		
6	6	12	16		5 Hours	20	233	11.65	13.40		
7	4	15	15		10 Hours	20	278	13.9	13.78		
8	7	11	9								
9	8	12	10								
10	16	9	12		ANOVA						
11	11	8	8		*Source of Variation*	*SS*	*df*	*MS*	*F*	*P-value*	*F crit*
12	9	10	19		Between Groups	135.63	2	67.82	4.45	0.02	3.16
13	6	7	16		Within Groups	868.10	57	15.23			
14	15	15	6								
15	13	14	19		Total	1003.73	59.0000				
16	11	17	17								
17	11	16	13								
18	7	20	18								
19	16	11	14								
20	19	14	15								
21	3	8	12								

Check Your Understanding

To check your understanding of the Anova: Single Factor tool, do the following two problems and check your answers in Figures A.19 and A.20 in Appendix A.

QS 45a. Compute the *F* value and evaluate it for significance for end-of-year math scores (out of 100 possible points) for three groups of homeschooled children in different enrichment programs. Use Data Set 45a.

QS 45b. Compute the *F* value and evaluate it for significance for four different levels of participation in a program that is designed to increase self-esteem on a measure of extroversion and introversion that ranges from 1 to 10, with 10 being high self-esteem. Use Data Set 45b.

EXCEL QUICKGUIDE 46

Anova: Two-Factor With Replication

What the Anova: Two-Factor With Replication Tool Does

The Anova: Two-Factor With Replication tests for differences between the means of two or more dependent groups or measures.

The Data Set

The data set used in this example is titled ANOVA WITH REPLICATION, and the question is, "Is there a difference in satisfaction level (SAT) over four training sessions (Training—Quarter 1 or TQ1, etc.) for 20 unemployed men and women undergoing quarterly training?"

Variable	*Description*
Gender	Male or female
TQ	Training score for each quarter
SAT	Level of satisfaction from 1 to 100 for each of four quarters

Using the Anova: Two-Factor With Replication Tool

1. Select the Data Analysis option from the Data tab.
2. Double-click on the Anova: Two-Factor With Replication option in the Data Analysis dialog box, and you will see the Anova: Two-Factor With Replication dialog box shown in Figure 46.1.
4. Click the RefEdit button in the Input Range entry box and select the data you want to use in the analysis. In this example, it is Cells B2 through F12. Click the RefEdit button.
5. Enter the number of rows in *each* sample in the Rows per sample box. In this case, it is 5.
6. Make sure the Alpha level is set at 0.05.
7. Click the RefEdit button and enter the Output Range. In this example, Cell A14 was selected. Click the RefEdit button again.
8. Click OK.

The Final Output

The results, shown in Figure 46.2, consist of several summary statistics for males and females and three F ratios testing the main effects of training, gender, and the interaction. With F values of 0.01 and 0.95, respectively, for Gender and the Gender × Training interaction, there was no significant outcome. For the main effect of training ($F = 4.98$), there was a significant outcome showing that regardless of gender, training changed over the four-quarter period. An examination of total means shows that there was an increase in level of satisfaction from TQ1 to TQ4, but it was not positive across all quarters.

Figure 46.1 The Anova: Two-Factor With Replication Dialog Box

Anova: Two-Factor With Replication
Input
Input Range:
Rows per sample:
Alpha: 0.05
OK
Cancel
Help
Output options
Output Range:
New Worksheet Ply:
New Workbook

Figure 46.2 The Anova: Two-Factor With Replication Output

	A	B	C	D	E	F	G
14	Anova: Two-Factor With Replication						
15							
16	SUMMARY	TQ1	TQ2	TQ3	TQ4	Total	
17	*Males*						
18	Count	5	5	5	5	20	
19	Sum	320	377	334	374	1405	
20	Average	64	75.4	66.8	74.8	70.25	
21	Variance	131	188.3	70.2	85.7	125.88	
22							
23	*Females*						
24	Count	5	5	5	5	20	
25	Sum	277	361	340	418	1396	
26	Average	55.4	72.2	68	83.6	69.8	
27	Variance	161.3	219.2	92	186.8	245.85	
28							
29	*Total*						
30	Count	10	10	10	10		
31	Sum	597	738	674	792		
32	Average	59.7	73.8	67.4	79.2		
33	Variance	150.46	183.96	72.49	142.62		
34							
35							
36	ANOVA						
37	*Source of Variation*	*SS*	*df*	*MS*	*F*	*P-value*	*F crit*
38	Sample	2.03	1	2.03	0.01	0.91	4.15
39	Columns	2119.28	3	706.43	4.98	0.01	2.90
40	Interaction	405.67	3	135.22	0.95	0.43	2.90
41	Within	4538.00	32	141.81			
42							
43	Total	7064.975	39				

Check Your Understanding

To check your understanding of the Anova: Two-Factor With Replication tool, do the following two problems and check your answers in Figures A.21 and A.22 in Appendix A.

QS 46a. Compute the F value and evaluate it for significance between different genders and three levels of program involvement (1-, 5-, and 10-hour weekly supplement) for history achievement on a test score ranging from 1 to 100. Use Data Set 46a.

QS 46b. Compute the F value and evaluate it for significance between two levels of participation in remedial work (high and low) and two levels of teaching programs on out-of-social school behaviors, as measured by the Out Of School (OOS) assessment tool. Scores on the OOS range from 1 to 10, with 10 being most social. Use Data Set 46b.

EXCEL QUICKGUIDE 47

Anova: Two-Factor Without Replication

What the Anova: Two-Factor Without Replication Tool Does

The Anova: Two-Factor Without Replication tests for differences between the means of two or more independent groups or measures.

The Data Set

The data set used in this example is titled ANOVA WITHOUT REPLICATION, and the question is, "Does happiness, as measured by the Happy Scale (HS), differ as a function of where people live (residence) and their political affiliation?"

Variable	*Description*
Residence	Rural or urban
Political Affiliation	Party 1 or Party 2
HS	Happiness Scale

Using the Anova: Two-Factor Without Replication Tool

1. Select the Data Analysis option from the Data tab.
2. Double-click on the Anova: Two-Factor Without Replication option in the Data Analysis dialog box, and you will see the Anova: Two-Factor Without Replication dialog box shown in Figure 47.1.
3. Click the RefEdit button in the Input Range entry box and select the data you want to use in the analysis. In this example, it is Cells B2 through D22. Click the RefEdit button again.
4. Be sure that the Labels box in the Input area is checked.
5. Be sure that the Alpha level is at 0.05.
6. Click the RefEdit button and enter the Output Range. In this example, Cell F1 was selected. Click the RefEdit button again.
7. Click OK.

The Final Output

The results, shown in Figure 47.2, are that the overall *F* value for Political Affiliation (0.82) is not significant and the value for Residence (4.41) is significant at the .05 level. There is no test of the interaction. The final conclusion is that rural residents (average happiness score of 7.3) are happier than urban residents (average happiness score of 5.9).

Figure 47.1 The Anova: Two-Factor Without Replication Dialog Box

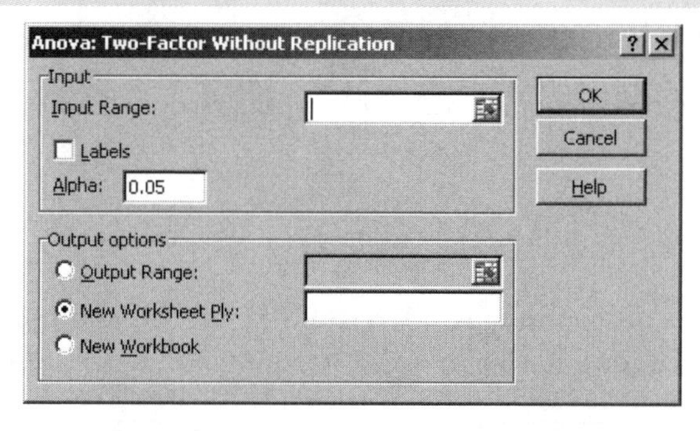

Figure 47.2 The Anova: Two-Factor Without Replication Output

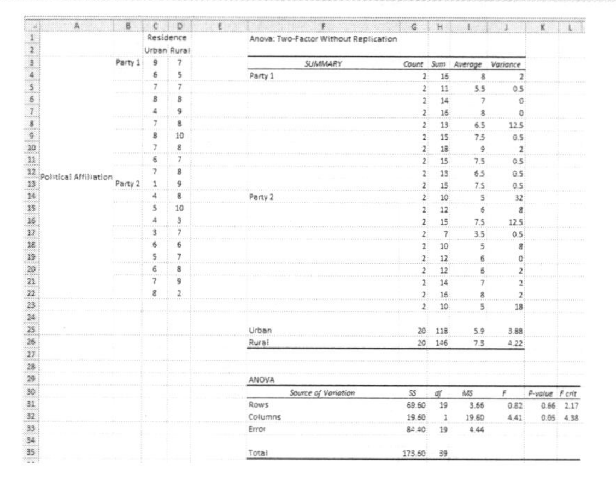

		Residence	
		Urban	Rural
	Party 1	9	7
		6	5
		7	7
		8	8
		4	9
		7	8
		8	10
		7	8
		6	7
Political Affiliation		7	8
	Party 2	1	9
		4	8
		5	10
		4	3
		3	7
		6	6
		5	7
		6	8
		7	9
		8	2

Anova: Two-Factor Without Replication

SUMMARY	*Count*	*Sum*	*Average*	*Variance*
Party 1	2	16	8	2
	2	11	5.5	0.5
	2	14	7	0
	2	16	8	0
	2	13	6.5	12.5
	2	15	7.5	0.5
	2	18	9	2
	2	15	7.5	0.5
	2	13	6.5	0.5
	2	15	7.5	0.5
Party 2	2	10	5	32
	2	12	6	8
	2	15	7.5	12.5
	2	7	3.5	0.5
	2	10	5	8
	2	12	6	0
	2	12	6	2
	2	14	7	2
	2	16	8	2
	2	10	5	18
Urban	20	118	5.9	3.88
Rural	20	146	7.3	4.22

ANOVA

Source of Variation	*SS*	*df*	*MS*	*F*	*P-value*	*F crit*
Rows	69.60	19	3.66	0.82	0.66	2.17
Columns	19.60	1	19.60	4.41	0.05	4.38
Error	84.40	19	4.44			
Total	173.60	39				

Check Your Understanding

To check your understanding of the ANOVA: Two Factor Without Replication tool, do the following two problems and check your answers in Figures A.23 and A.24 in Appendix A.

QS 47a. Compute the *F* value and evaluate it for significance among three different speaking programs with three different levels of experience (1, 2, and 3) with the outcome variable being number of speeches per year. Use Data Set 47a.

QS 47b. Compute the *F* value and evaluate it for significance between different neighborhoods in two communities and two levels of advertising on recycling rates measured from 0 to 100, with 100 representing "recycles as much as possible." Use Data Set 47b.

EXCEL QUICKGUIDE 48

The Correlation Tool

What the Correlation Tool Does

The Correlation tool computes the value of the Pearson product-moment correlation between two variables.

The Data Set

The data set used in this example is titled CORRELATION, and the question is, "What is the correlation between number of years teaching and teaching skills?"

Variable	*Description*
Years Teaching	Number of years teaching
Teaching Skills	Teaching skills rated from 1 to 10

Using the Correlation Tool

1. Select the Data Analysis option from the Data tab.
2. Double-click on the Correlation option in the Data Analysis dialog box, and you will see the Correlation dialog box shown in Figure 48.1.
3. Click the RefEdit button in the Input Range entry box and select the data you want to use in the analysis. In this example, it is Cells C1 through D21. Click the RefEdit button again.
4. Be sure to check the Labels in First Row box.
5. Click the RefEdit button and enter the Output Range. In this example, Cell F1 is selected. Click the RefEdit button again.
6. Click OK.

The Final Output

The result, shown in Figure 48.2, is a correlation between Years Teaching and Teaching Skills of .67.

Figure 48.1 The Correlation Dialog Box

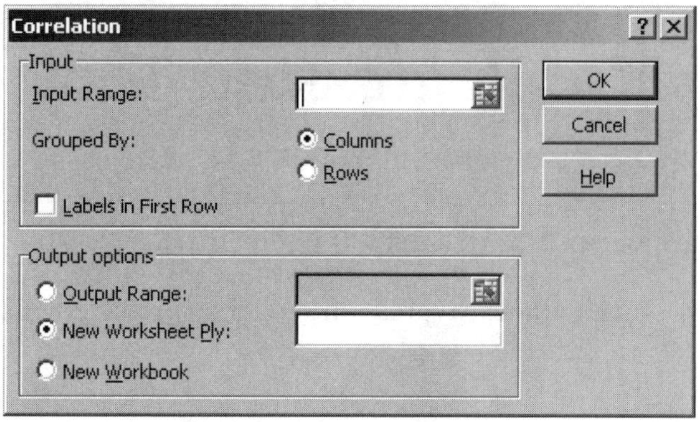

Correlation

Input

Input Range:

Grouped By: Columns / Rows

Labels in First Row

Output options

Output Range:

New Worksheet Ply:

New Workbook

OK

Cancel

Help

Figure 48.2 The Correlation Output

	A	B	C	D	E	F	G	H
1		ID	Years Teaching	Teaching Skills			*Years Teaching*	*Teaching Skills*
2		1	14	8		Years Teaching	1	
3		2	21	10		Teaching Skills	0.67	1
4		3	5	6				
5		4	26	9				
6		5	13	8				
7		6	9	6				
8		7	11	9				
9		8	5	5				
10		9	17	9				
11		10	25	9				
12		11	31	9				
13		12	19	7				
14		13	12	9				
15		14	5	8				
16		15	1	3				
17		16	11	7				
18		17	16	8				
19		18	13	9				
20		19	15	9				
21		20	7	8				

Check Your Understanding

To check your understanding of the Correlation tool, do the following two problems and check your answers in Figures A.25 and A.26 in Appendix A.

QS 48a. Compute the correlation between consumption of ice cream (times eaten) and crime rate (frequency) in 15 Midwestern cities. Use Data Set 48a.

QS 48b. Compute the correlation between speed of completing a task and the accuracy with which that task is completed. Use Data Set 48b.

EXCEL QUICKGUIDE 49

The Regression Tool

What the Regression Tool Does

The Regression tool uses a linear regression model to predict a *Y* outcome from an *X* variable.

The Data Set

The data set used in this example is titled REGRESSION, and the question is, "How well does the average number of hours spent studying predict GPA?"

Variable	*Description*
Hours	Hours of studying each week
GPA	Grade point average

Using the Regression Tool

1. Select the Data Analysis option from the Data tab.
2. Double-click on the Regression option in the Data Analysis dialog box, and you will see the Regression dialog box shown in Figure 49.1.
3. Click the RefEdit button in the Input Y Range entry box and select the data you want to use in the analysis. In this example, it is Cells B1 through B21. Click the RefEdit button again.
4. Repeat Step 3 for the Input X Range entry box. In this example, select Cells A1 through A21. Click the RefEdit button again.
5. Be sure to check the Labels box.
6. Check the Confidence Level box and make sure the confidence level is set to 95%.
7. Click the Output Range button, click the RefEdit button, and enter the output range. In this example, Cell D1 is selected. Click the RefEdit button again.
8. Click OK.

The Final Output

The results in Figure 49.2 show the formula for the regression line to be $Y' = 0.05x + 1.64$.

Figure 49.1 The Regression Dialog Box

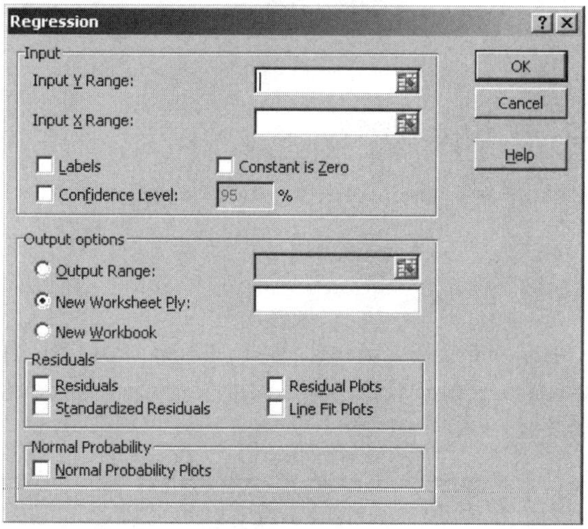

Regression

Input

Input Y Range:

Input X Range:

Labels

Constant is Zero

Confidence Level: 95 %

OK

Cancel

Help

Output options

Output Range:

New Worksheet Ply:

New Workbook

Residuals

Residuals

Residual Plots

Standardized Residuals

Line Fit Plots

Normal Probability

Normal Probability Plots

Figure 49.2 The Regression Output

	A	B	C	D	E	F	G	H	I	J	K	L
1	Hours (X)	GPA (Y)		SUMMARY OUTPUT								
2	15	3.4										
3	11	2.1		*Regression Statistics*								
4	21	2.8		Multiple R	0.50							
5	25	3.7		R Square	0.25							
6	24	2.1		Adjusted R Square	0.21							
7	17	1.8		Standard Error	0.64							
8	9	1.3		Observations	20							
9	25	3.2										
10	21	3.7		ANOVA								
11	22	2.6			*df*	*SS*	*MS*	*F*	*Significance F*			
12	18	2.1		Regression	1	2.55	2.55	6.14	0.02			
13	13	2.9		Residual	18	7.49	0.42					
14	27	3.1		Total	19	10.04						
15	16	1.9										
16	31	3.3			*Coefficients*	*Standard Error*	*t Stat*	*P-value*	*Lower 95%*	*Upper 95%*	*Lower 95.0%*	*Upper 95.0%*
17	36	3.8		Intercept	1.64	0.45	3.63	0.00	0.69	2.59	0.69	2.59
18	22	2.8		Hours (X)	0.05	0.02	2.48	0.02	0.01	0.10	0.01	0.10
19	23	1.8										
20	21	2.5										
21	9	3.1										

Check Your Understanding

To check your understanding of the Regression tool, do the following two problems and check your answers in Figures A.27 and A.28 in Appendix A.

QS 49a. Compute the regression line for predicting wins from number of injuries teams suffer. Use Data Set 49a.

QS 49b. Compute the regression line for predicting number of cars sold in a year from salesperson's years of experience. Use Data Set 49b.

EXCEL QUICKGUIDE 50

The Histogram Tool

What the Histogram Tool Does

The Histogram tool creates an image of the frequencies of set values organized in classes.

The Data Set

The data set used in this example is titled HISTOGRAM, and the question is, "What is the frequency of first-year, second-year, third-year, and fourth-year students in a sample of 25 students?"

Variable	*Description*
Class	1 = 1st year, 2 = 2nd year, 3 = 3rd year, and 4 = 4th year
Bin	The group assignment criteria (Class = 1–4)

Using the Histogram Tool

1. Select the Data Analysis option from the Data tab.
2. Double-click on the Histogram option in the Data Analysis dialog box, and you will see the Histogram dialog box shown in Figure 50.1.
3. Click the RefEdit button in the Input Range entry box and select the data you want to use in the analysis. In this example, it is Cells A1 through A26. Click the RefEdit button.
4. Click the RefEdit button in the Bin Range entry box and select the data you want to use in the analysis. In this example, it is Cells B1 through B5. Click the RefEdit button again.
5. Be sure to check the Labels box.
6. Click the Output Range button, click the RefEdit button, and enter the output range. In this example, Cell D1 is selected. Click the RefEdit button again.
7. Click the Cumulative Percentage and Chart Output boxes.
8. Click OK.

The Final Output

The results in Figure 50.2 show the frequency of each value in the Bin Range plus the cumulative frequency. A histogram of the frequencies and a line showing the cumulative frequencies are also produced.

Figure 50.1 The Histogram Dialog Box

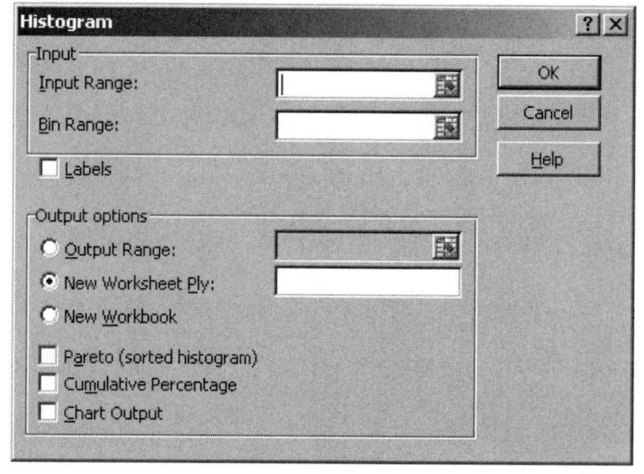

Figure 50.2 The Histogram Output

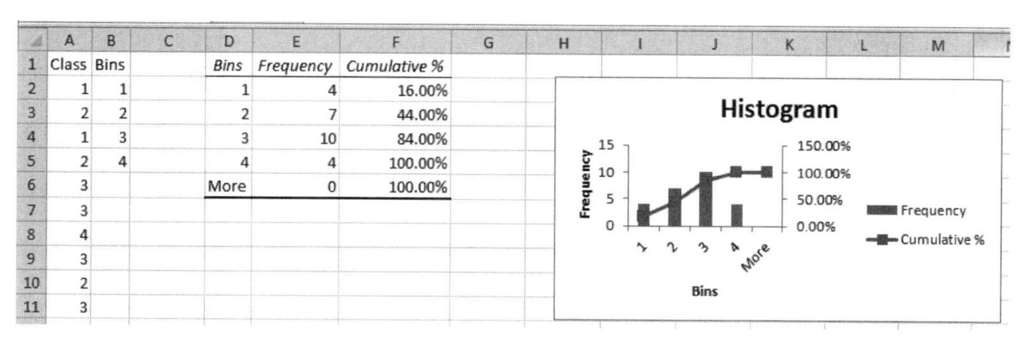

	A	B	C	D	E	F
1	Class	Bins		*Bins*	*Frequency*	*Cumulative %*
2	1	1		1	4	16.00%
3	2	2		2	7	44.00%
4	1	3		3	10	84.00%
5	2	4		4	4	100.00%
6	3			More	0	100.00%
7	3					
8	4					
9	3					
10	2					
11	3					

Check Your Understanding

To check your understanding of the Histogram tool, do the following two problems and check your answers in Figures A.29 and A.30 in Appendix A.

QS 50a. Generate a histogram for choice of meal (1 = vegan, 2 = vegetarian, 3 = meat) for 100 diners. Use Data Set 50a.

QS 50b. Generate a histogram for favorite type of recreation (1 = walking, 2 = running, 3 = swimming, 4 = yoga) for 200 community members. Use Data Set 50b.

Appendix A

Answers to QuickGuide Questions

Part I: Using Excel Functions

QS 1a. The average number of visitors is 5,071.

QS 1b. The average age is 23.35 years.

QS 2a. The median price of a home across the 10 communities is $324,544.

QS 2b. The median height is 74.50 inches.

QS 3a. The preferred topping is 1 or hot fudge.

QS 3b. The modal age range is 2 or between 21 and 34 years of age.

QS 4a. The geometric mean for the average achievement score is .79 or 79%.

QS 4b. The geometric mean for the average amount of groceries purchased is $95.32.

QS 5a. The standard deviation for the percent correct on the spelling tests is 0.17.

QS 5b. The standard deviation for the number of adults who received a flu shot is 1,285.7.

QS 6a. The standard deviation for the number of users of the community center is 134.9.

QS 6b. The standard deviation for the number of shoppers at Hobbs toy store is 68.18.

QS 7a. The variance for the sample of math test scores is 217.73.

QS 7b. The variance for the sample of aggressive behavior scores is 2.77.

QS 8a. The variance for the population of annual income across the 10 cities is $236,085,026.

QS 8b. The variance for the population of BMI scores is 9.93.

QS 9a. Frequency distributions for Condition:

1: 21

2: 39

QS 9b. Frequency distributions for Consumer Response:

1: 7

2: 15

3: 18

QS 10a. Cumulative probabilities for sales by month:

Month	*Sales*	*NORMDIST (Cumulative Probability)*
January	$25,456	22.06%
February	$25,465	22.06%
March	$44,645	32.09%
April	$54,345	37.79%
May	$55,456	38.47%
June	$56,456	39.07%
July	$65,456	44.66%
August	$66,787	45.50%
September	$76,564	51.69%
October	$87,656	58.65%
November	$254,657	99.80%
December		
Sales Mean: $73,903.91		
Sales SD: $62,902.15		

QS 10b. Cumulative probabilities for number of cars in inventory:

Quarter	*Number*	*NORMDIST (Cumulative Probability)*
3rd Quarter	234	19.55%
1st Quarter	245	22.94%
2nd Quarter	346	63.00%
4th Quarter	434	89.74%
Quarter Mean: 314.75		
Quarter SD: 94.13		

QS 11a. The 50th percentile for use of library facilities is 248.5.

QS 11b. The 80th percentile for number of meals prepared is 333.6.

QS 12a. The percent rank for a raw score of 89 is 79%.

QS 12b. The percent rank for a house priced at $156,500 is 0%.

QS 13a. The value of the 2nd quartile is 16.

QS 13b. The value of the 3rd quartile is $563,500.

QS 14a. A score of 57 is a rank of 41th.

QS 14b. Team #6 ranked 14th.

QS 15a. Standardized scores:

ID	*Test Score*	*STANDARDIZE*
1	75	–1.36
2	80	–0.75
3	85	–0.14
4	87	0.11
5	74	–1.49
6	96	1.21
7	95	1.09
8	73	–1.61
9	90	0.48
10	88	0.23

(Continued)

(Continued)

ID	*Test Score*	*STANDARDIZE*
11	95	1.09
12	87	0.11
13	89	0.35
14	85	−0.14
15	84	−0.26
16	93	0.84
17	90	0.48
18	74	−1.49
19	77	−1.12
20	76	−1.24
21	96	1.21
22	86	−0.01
23	81	−0.63
24	98	1.46
25	99	1.58
AVERAGE	86.12	
STDEV	8.16	

QS 15b. The corresponding standardized score for a raw score of 76 is −1.11.

QS 16a. The covariance between time to first response and correct responses is 5.58.

QS 16b. The covariance between study time and GPA is −.31.

QS 17a. The correlation between children retained and parental involvement is −.36.

QS 17b. The correlation between the number of water treatment plants and incidence of disease is −.23.

QS 18a. The correlation between between years in teaching and teaching evaluations from high school students is .15.

QS 18b. The correlation between ice cream consumption and crime rate is .31.

QS 19a. The intercept is $50,435.78.

QS 19b. The intercept is 7.45.

QS 20a. The slope is $7,134.20.

QS 20b. The slope is –.04.

QS 21a. Trend for new income scores:

$74,629.06

$63,990.76

$66,844.44

$70,411.54

$76,832.32

QS 21b. Trend for new success scores:

7.2

7.3

6.9

7.2

QS 22a. Forecast for new losses:

4 losses

5 losses

10 losses

7 losses

8 losses

QS 22b. Forecast for new history test scores:

80

82

85

QS 23a. The RSQ between height and weight is .4.

QS 23b. The RSQ between quality of care and decline in infant mortality is .8.

QS 24a. The value of CHISQ.DIST is .28.

QS 24b. The value of CHISQ.DIST is .02.

QS 25a. The probability is .75 that the proportion of actual and expected community center users differ.

QS 25b. The probability is .06 that people have a preference for a particular type of detergent.

QS 26a. The F.DIST value is .96.

QS 26b. The F.DIST value is .83.

QS 27a. The F.TEST value is .12.

QS 27b. The F.TEST value is .18.

QS 28a. The T.DIST value is .62.

QS 28b. The T.DIST value is .97.

QS 29a. The probability associated with a *t*-test for the difference between test scores is 2.86-12.

QS 29b. The probability associated with a *t*-test for the difference between ratings for two different dishes is .20.

QS 30a. The probability that a math test score of 83 is characteristic of the entire set of scores is .98.

QS 30b. The probability that the free-throw shooting percentage of 63 is unique is .05.

QS 31a. The AVERAGEIF sales value for those members of the A Team is $520,982.

QS 31b. The AVERAGEIF incidence for infections rate (per 1,000 patients) in hospitals in communities with fewer than 200,000 people is 5.63.

QS 32a. The number of people who purchased insurance is 6.

QS 32b. The number of people who gave their age and years of service is 9 and 14, respectively.

QS 33a. The number of people who decided to enroll is 13.

QS 33b. The number of communities that adopted curbside or alley recycling is 15.

QS 34a. The number of students from a group of 35 who did not show up in an elective class is 10.

QS 34b. The number of communities that do not have an adequate tax base is 6.

QS 35a. The number of students who were admitted to the program and subsequently enrolled is 16.

QS 35b. The number of businesses that are open on the weekend is 17.

Part II: Using the Analysis ToolPak

Figure A.1 Descriptive Statistics for Sales Tax Rates

	A	B	C	D	E
1	City	Tax Rate		*Tax Rate*	
2	1	3.0%			
3	2	3.5%		Mean	0.057
4	3	6.0%		Standard Error	0.007
5	4	7.0%		Median	0.058
6	5	4.8%		Mode	0.070
7	6	6.7%		Standard Deviation	0.021
8	7	9.8%		Sample Variance	0.000
9	8	3.7%		Kurtosis	0.271
10	9	5.6%		Skewness	0.542
11	10	7.0%		Range	0.068
12				Minimum	0.030
13				Maximum	0.098
14				Sum	0.571
15				Count	10

Figure A.2 Descriptive Statistics for Car Sales

L25 *fx*

	A	B	C	D	E
1	Week	Sales		*Sales*	
2	1	45			
3	2	41		Mean	49.75
4	3	47		Standard Error	5.56
5	4	66		Median	46.00
6				Mode	#N/A
7				Standard Deviation	11.12
8				Sample Variance	123.58
9				Kurtosis	3.12
10				Skewness	1.70
11				Range	25.00
12				Minimum	41.00
13				Maximum	66.00
14				Sum	199.00
15				Count	4

Figure A.3 Moving Average for Number of Parking Spaces Used

	A	B
1	Number of Spaces Used	MOVING AVERAGE
2	80	#N/A
3	82	#N/A
4	100	#N/A
5	91	88.75
6	82	88.50
7	81	87.75
8	97	89.50
9	98	90.75
10	87	94.00
11	94	90.00
12	81	89.75
13	97	88.00
14	80	85.75
15	85	87.00
16	86	87.50
17	99	89.50
18	88	91.25
19	92	93.25
20	94	91.00
21	90	92.50
22	94	90.25
23	83	87.00
24	81	85.75
25	85	83.00

Figure A.4 Moving Average for Hours of Training

	A	B	C
1	Weeks	Hours Training	MOVING AVERAGE
2	1	16	#N/A
3	2	19	#N/A
4	3	19	#N/A
5	4	16	#N/A
6	5	16	17.2
7	6	18	17.6
8	7	15	16.8
9	8	15	16
10	9	17	16.2
11	10	17	16.4
12	11	16	16
13	12	19	16.8

Figure A.5 A Set of 25 Random Numbers

	A	B
1	ID	RANDOM NUMBER
2	1	0.301326
3	2	1.469444
4	3	0.385979
5	4	0.882034
6	5	1.167742
7	6	0.11819
8	7	-0.48372
9	8	1.769284
10	9	0.543838
11	10	3.150755
12	11	1.394107
13	12	1.238587
14	13	1.640471
15	14	0.912181
16	15	-0.20291
17	16	0.35558
18	17	0.52174
19	18	2.042613
20	19	-0.35595
21	20	0.961167
22	21	1.218954
23	22	-0.44202
24	23	1.545328
25	24	1.201123
26	25	2.905119

Figure A.6 Random Numbers Associated With 10 Participants

	A	B
1	Participant	RANDOM NUMBER
2	1	0.9
3	2	1.6
4	3	1.2
5	4	0.7
6	5	1.8
7	6	1.4
8	7	0.1
9	8	0.5
10	9	1.6
11	10	1.3

Figure A.7 Rank and Percentile Scores for Salary

	A	B	C	D	E
1	Salary	*Point*	*Salary*	*Rank*	*Percent*
2	$ 93,110	3	$97,935	1	100.00%
3	$ 88,794	9	$93,287	2	94.70%
4	$ 97,935	1	$93,110	3	89.40%
5	$ 81,601	17	$92,234	4	84.20%
6	$ 72,021	7	$91,416	5	78.90%
7	$ 83,162	2	$88,794	6	73.60%
8	$ 91,416	6	$83,162	7	68.40%
9	$ 55,987	20	$82,651	8	63.10%
10	$ 93,287	4	$81,601	9	57.80%
11	$ 57,681	12	$77,665	10	52.60%
12	$ 62,769	18	$77,108	11	47.30%
13	$ 77,665	5	$72,021	12	42.10%
14	$ 64,168	13	$64,168	13	36.80%
15	$ 57,915	16	$63,876	14	31.50%
16	$ 51,383	11	$62,769	15	26.30%
17	$ 63,876	14	$57,915	16	21.00%
18	$ 92,234	10	$57,681	17	15.70%
19	$ 77,108	19	$57,616	18	10.50%
20	$ 57,616	8	$55,987	19	5.20%
21	$ 82,651	15	$51,383	20	0.00%

Figure A.8 Rank and Percentile Scores for Hours of Studying

	A	B	C	D	E
1	Hours Studying	*Point*	*Hours Studying*	*Rank*	*Percent*
2	16	8	32	1	100.00%
3	21	14	31	2	92.80%
4	24	11	29	3	85.70%
5	25	15	28	4	78.50%
6	26	5	26	5	71.40%
7	12	4	25	6	64.20%
8	8	3	24	7	57.10%
9	32	13	22	8	50.00%
10	14	2	21	9	42.80%
11	17	10	17	10	35.70%
12	29	1	16	11	21.40%
13	16	12	16	11	21.40%
14	22	9	14	13	14.20%
15	31	6	12	14	7.10%
16	28	7	8	15	0.00%

Figure A.9 A Random Selection of 15 Participants

	A	B	C	D	E	F	G	H	I	J	K	L
1	ID											Sample
2	18	31	71	25	92	34	10	14	16	26		67
3	62	2	79	67	19	53	86	55	57	49		29
4	7	65	39	44	48	56	10	55	32	78		53
5	26	43	63	19	22	57	31	56	86	80		25
6	59	92	86	96	25	13	6	27	37	4		97
7	30	9	11	12	28	42	69	65	18	87		95
8	95	16	97	72	100	76	19	56	45	29		32
9	5	84	92	22	92	36	74	13	43	80		97
10	23	74	43	35	74	34	38	81	21	97		96
11												44
12												74
13												4
14												96
15												31
16												56

Figure A.10 A Random Selection of a Sample of 20

	A	B	C	D	E	F	G	H	I	J	K	L
1	ID											Sample
2	99	158	26	62	149	101	133	131	195	101		129
3	143	159	126	96	20	174	99	109	32	40		13
4	63	173	25	180	38	109	193	162	62	114		51
5	133	49	48	155	140	30	164	133	172	10		88
6	148	61	198	122	14	182	194	168	45	102		132
7	56	39	22	187	30	164	191	48	78	195		39
8	30	95	137	190	66	133	13	107	141	73		67
9	102	121	139	142	29	183	190	76	192	107		98
10	148	64	171	135	14	145	3	24	43	157		89
11	101	145	26	114	16	5	106	106	131	157		68
12	159	167	192	64	190	144	187	183	77	85		125
13	67	113	133	141	7	148	92	168	73	164		118
14	2	27	119	187	45	98	153	24	62	133		145
15	33	178	115	143	101	76	3	171	30	156		87
16	179	65	48	188	118	60	144	97	184	47		100
17	192	175	161	104	110	137	55	19	14	179		149
18	154	182	59	95	132	65	89	114	147	76		173
19	8	132	78	137	143	185	122	165	92	23		62
20	1	123	25	184	34	94	72	151	193	7		30
21	143	92	137	85	81	107	115	126	25	59		92

Figure A.11 The z Value of −1.51 Indicates That There Is No Significant Difference Between Groups of Homeowners

	A	B	C	D	E	F	G
1		Homeowners 1	Homeowners 2		z-Test: Two Sample for Means		
2		3	4				
3		5	3			*Homeowners 1*	*Homeowners 2*
4		1	3		Mean	3.50	3.90
5		1	4		Known Variance	1.51	0.79
6		2	5		Observations	20	20
7		3	5		Hypothesized Mean Difference	0.00	
8		2	5		z	-1.18	
9		4	4		P(Z<=z) one-tail	0.12	
10		5	3		z Critical one-tail	1.64	
11		5	3		P(Z<=z) two-tail	0.24	
12		4	2		z Critical two-tail	1.96	
13		4	3				
14		3	4				
15		3	5				
16		4	5				
17		4	4				
18		5	5				
19		4	4				
20		3	4				
21		5	3				
22							
23	VAR.P	1.507	0.787				

Figure A.12 The z Value of 1.80 Indicates That There Is a Significant Difference Between the Two Groups

J28

	A	B	C	D	E	F
1		Group 1	Group 2	z-Test: Two Sample for Means		
2		139	96			
3		134	90		*Group 1*	*Group 2*
4		125	113	Mean	121.36	110.86
5		120	101	Known Variance	308.52	170.27
6		115	104	Observations	14	14
7		91	123	Hypothesized Mean Difference	0	
8		129	106	z	1.80	
9		141	106	P(Z<=z) one-tail	0.04	
10		122	107	z Critical one-tail	1.64	
11		154	139	P(Z<=z) two-tail	0.07	
12		121	130	z Critical two-tail	1.96	
13		117	107			
14		94	106			
15		97	124			
16						
17	VAR.P	308.515	170.265			

Figure A.13 The t Value of −1.87 Indicates That There Is a Significant Difference Between Before and After Scores

	A	B	C	D	E
1	Before Course	After Course	t-Test: Paired Two Sample for Means		
2	144	139			
3	144	145		*Before Course*	*After Course*
4	145	146	Mean	142.40	145.64
5	141	149	Variance	4.75	56.49
6	145	139	Observations	25	25
7	141	147	Pearson Correlation	-0.06	
8	142	143	Hypothesized Mean Difference	0.00	
9	143	166	df	24	
10	142	151	t Stat	-2.04	
11	145	144	P(T<=t) one-tail	0.03	
12	140	140	t Critical one-tail	1.71	
13	145	144	P(T<=t) two-tail	0.05	
14	140	136	t Critical two-tail	2.06	
15	141	140			
16	140	137			
17	142	149			
18	140	147			
19	140	167			
20	145	146			
21	144	145			
22	145	145			
23	144	137			
24	144	150			
25	140	144			
26	138	145			

Figure A.14 The *t* Value of .86 Indicates That the BMI Score Did Not Change as a Function of the Treatment

	A	B	C	D	E
1	BMI Before	BMI After	t-Test: Paired Two Sample for Means		
2	29	24			
3	25	26		*BMI Before*	*BMI After*
4	24	23	Mean	25.90	25.30
5	35	27	Variance	12.52	5.48
6	22	25	Observations	20	20
7	28	27	Pearson Correlation	0.50	
8	26	24	Hypothesized Mean Difference	0	
9	27	25	df	19	
10	26	24	t Stat	0.86	
11	23	22	P(T<=t) one-tail	0.20	
12	21	23	t Critical one-tail	1.73	
13	21	25	P(T<=t) two-tail	0.40	
14	24	23	t Critical two-tail	2.09	
15	25	25			
16	24	26			
17	32	30			
18	26	22			
19	30	28			
20	25	27			
21	25	30			

Figure A.15 The *t* Value of −2.59 Indicates That There Is a Significant Difference Between the Two Groups

I26 *fx*

	A	B	C	D	E
1	Group 1	Group 2	t-Test: Two-Sample Assuming Unequal Variances		
2	9.8	6.5			
3	8.7	7.6		*Group 1*	*Group 2*
4	3.2	9.1	Mean	6.05	9.8
5	5.6	13.1	Variance	11.43	9.50
6	4.7	16.5	Observations	10	10
7	2	9.4	Hypothesized Mean Difference	0	
8	2.1	8	df	18	
9	6.6	9.8	t Stat	-2.59	
10	5.5	6.9	P(T<=t) one-tail	0.01	
11	12.3	11.1	t Critical one-tail	1.73	
12			P(T<=t) two-tail	0.02	
13			t Critical two-tail	2.10	

Figure A.16 The t Value of 2.79 Indicates That There Is a Significant Difference Between the Two Groups

	A	B	C	D	E
1	Public Debt	Private Debt	t-Test: Two-Sample Assuming Unequal Variances		
2	$ 13,267	$ 8,573			
3	$ 12,562	$ 10,150		*Public Debt*	*Private Debt*
4	$ 18,995	$ 10,607	Mean	$ 15,303.70	$ 10,607.50
5	$ 14,996	$ 10,531	Variance	$ 6,351,032.01	$ 30,830,306.27
6	$ 17,181	$ 7,572	Observations	10	14
7	$ 14,488	$ 7,103	Hypothesized Mean Difference	0	
8	$ 14,760	$ 8,497	df	19	
9	$ 18,591	$ 9,240	t Stat	2.79	
10	$ 16,704	$ 10,919	P(T<=t) one-tail	0.01	
11	$ 11,493	$ 9,770	t Critical one-tail	1.73	
12		$ 21,345	P(T<=t) two-tail	0.01	
13		$ 24,312	t Critical two-tail	2.09	
14		$ 5,463			
15		$ 4,423			

Figure A.17 The t Value of 2.50 Indicates That There Is a Significant Difference Between the Cessation and Control Groups

	A	B	C	D	E	F
1	ID	Cessation	Control	t-Test: Two-Sample Assuming Equal Variances		
2	1	24	7			
3	2	21	12		*Cessation*	*Control*
4	3	23	14	Mean	31.95	22.15
5	4	24	15	Variance	181.52	142.31
6	5	32	21	Observations	20	13
7	6	45	33	Pooled Variance	166.34	
8	7	42	21	Hypothesized Mean Difference	0.00	
9	8	36	34	df	31	
10	9	52	32	t Stat	2.13	
11	10	12	3	P(T<=t) one-tail	0.02	
12	11	25	21	t Critical one-tail	1.70	
13	12	42	32	P(T<=t) two-tail	0.04	
14	13	57	43	t Critical two-tail	2.04	
15	14	43				
16	15	44				
17	16	23				
18	17	15				
19	18	44				
20	19	23				
21	20	12				

Figure A.18 The *t* Value of –.18 Indicates There Is No Significant Difference Between the Brick-and-Mortar and the Online Programs

	A	B	C	D	E	F
1	ID	Bricks	Online	t-Test: Two-Sample Assuming Equal Variances		
2	1	$ 21,714	$ 21,992			
3	2	$ 20,309	$ 23,774		*Bricks*	*Online*
4	3	$ 22,819	$ 20,408	Mean	$ 24,429.92	$ 25,176.16
5	4	$ 21,370	$ 20,472	Variance	$ 9,931,818.91	$ 10,330,008.47
6	5	$ 25,595	$ 28,110	Observations	25	25
7	6	$ 23,995	$ 20,055	Pooled Variance	$ 10,130,913.69	
8	7	$ 29,876	$ 26,025	Hypothesized Mean Difference	0.00	
9	8	$ 27,677	$ 27,942	df	48.00	
10	9	$ 20,606	$ 24,014	t Stat	-0.83	
11	10	$ 28,850	$ 24,072	P(T<=t) one-tail	0.21	
12	11	$ 25,853	$ 29,150	t Critical one-tail	1.68	
13	12	$ 29,714	$ 21,294	P(T<=t) two-tail	0.41	
14	13	$ 20,537	$ 29,135	t Critical two-tail	2.01	
15	14	$ 28,800	$ 27,912			
16	15	$ 23,827	$ 21,160			
17	16	$ 24,379	$ 27,607			
18	17	$ 20,701	$ 24,776			
19	18	$ 23,803	$ 27,118			
20	19	$ 24,978	$ 28,089			
21	20	$ 21,522	$ 25,846			
22	21	$ 27,768	$ 29,369			
23	22	$ 22,363	$ 20,550			
24	23	$ 22,605	$ 26,375			
25	24	$ 22,603	$ 25,007			
26	25	$ 28,484	$ 29,152			

Figure A.19 The *F* Value of .24 Indicates That There Is No Significant Difference in Math Performance Among the Three Groups, With Homeschoolers, 3, Having the Highest Average

N29 *fx*

	A	B	C	D	E	F	G	H	I	J	K
1	Home School 1	Home School 2	Home School 3		Anova: Single Factor						
2	91	79	62								
3	72	81	100		SUMMARY						
4	97	86	67		*Groups*	*Count*	*Sum*	*Average*	*Variance*		
5	84	86	92		Home School 1	25	1959	78.36	120.24		
6	67	90	94		Home School 2	25	2000	80	134.3		
7	78	70	82		Home School 3	25	2014	80.56	156.2		
8	82	60	83								
9	63	83	70								
10	73	90	89		ANOVA						
11	92	92	97		*Source of Variation*	*SS*	*df*	*MS*	*F*	*P-value*	*F crit*
12	66	78	70		Between Groups	65.36	2	32.68	0.24	0.79	3.12
13	76	94	60		Within Groups	9857.92	72	136.92			
14	66	64	61								
15	65	90	97		Total	9923.28	74				
16	86	83	78								
17	68	97	81								
18	69	98	98								
19	91	82	76								
20	65	86	82								
21	91	64	85								
22	93	83	85								
23	70	70	77								
24	89	64	60								
25	78	63	79								
26	87	67	89								

Figure A.20 The *F* Value of 89.18 Indicates That There Is a Significant Difference Among Self-Esteem Programs, With Self-Esteem Program 3 Having the Highest Average

	A	B	C	D	E	F	G	H	I	J	K
1	Self Esteem 1	Self Esteem 2	Self Esteem 3	Self Esteem 4	Anova: Single Factor						
2	2	5	10	5							
3	1	3	8	4	SUMMARY						
4	5	3	9	2	*Groups*	*Count*	*Sum*	*Average*	*Variance*		
5	2	4	10	3	Self Esteem 1	20	61	3.05	2.79		
6	1	6	10	4	Self Esteem 2	20	101	5.05	2.26		
7	2	7	10	3	Self Esteem 3	20	182	9.10	0.62		
8	1	4	9	2	Self Esteem 4	20	62	3.10	1.57		
9	5	7	9	2							
10	1	6	10	5							
11	4	5	8	4	ANOVA						
12	2	5	9	4	*Source of Variation*	*SS*	*df*	*MS*	*F*	*P-value*	*F crit*
13	1	3	9	2	Between Groups	484.05	3	161.35	89.18	0.00	2.72
14	3	7	8	3	Within Groups	137.50	76	1.81			
15	3	7	9	5							
16	5	6	10	2	Total	621.55	79				
17	5	4	9	3							
18	5	5	8	1							
19	5	3	8	4							
20	3	4	10	1							
21	5	7	9	3							

Figure A.21 The *F* Values Show a Main Effect for Gender and No Effect for Level but a Significant Interaction Effect

	A	B	C	D	E	F	G
14	Anova: Two-Factor With Replication						
15							
16	SUMMARY	Level 1	Level 1	Level 2	Total		
17	*Males*						
18	Count	5	5	5	15		
19	Sum	355	397	421	1173		
20	Average	71	79.4	84.2	78.2		
21	Variance	53.5	8.8	93.2	76.31428571		
22							
23	*Females*						
24	Count	5	5	5	15		
25	Sum	353	402	416	1171		
26	Average	70.6	80.4	83.2	78.07		
27	Variance	61.3	9.8	130.7	88.92		
28							
29	*Total*						
30	Count	10	10	10			
31	Sum	708	799	837			
32	Average	70.8	79.9	83.7			
33	Variance	51.07	8.54	99.79			
34							
35							
36	ANOVA						
37	*Source of Variation*	*SS*	*df*	*MS*	*F*	*P-value*	*F crit*
38	Sample	0.13	1	0.13	0.00	0.96	4.26
39	Columns	878.87	2	439.43	7.38	0.00	3.40
40	Interaction	5.27	2	2.63	0.04	0.96	3.40
41	Within	1429.20	24	59.55			
42							
43	Total	2313.47	29				

Figure A.22 The *F* Values Show a Main Effect for Participation but No Main Effect for Teaching or an Interaction Effect

	A	B	C	D	E	F	G
14	Anova: Two-Factor With Replication						
15							
16	SUMMARY	In School	In Home	Total			
17	*High*						
18	Count	5	5	10			
19	Sum	24	24	48			
20	Average	4.8	4.8	4.8			
21	Variance	1.7	0.7	1.07			
22							
23	*Low*						
24	Count	5	5	10			
25	Sum	27	33	60			
26	Average	5.4	6.6	6			
27	Variance	0.8	0.3	0.89			
28							
29	*Total*						
30	Count	10	10				
31	Sum	51	57				
32	Average	5.1	5.7				
33	Variance	1.21	1.34				
34							
35							
36	ANOVA						
37	*Source of Variation*	*SS*	*df*	*MS*	*F*	*P-value*	*F crit*
38	Sample	7.2	1	7.2	8.23	0.01	4.49
39	Columns	1.8	1	1.8	2.06	0.17	4.49
40	Interaction	1.8	1	1.8	2.06	0.17	4.49
41	Within	14	16	0.88			
42							
43	Total	24.8	19				

Figure A.23 The *F* Values Show No Main Effect for Experience but a Significant Main Effect for Type of Speaking Program

ANOVA						
Source of Variation	*SS*	*df*	*MS*	*F*	*P-value*	*F crit*
Rows	91.66	29	3.16	1.22	0.26	1.66
Columns	60.09	2	30.04	11.57	0.00	3.16
Error	150.58	58	2.60			
Total	302.32	89				

Figure A.24 The F Values Show a Significant Main Effect for Neither Community nor Advertising Rates

ANOVA						
Source of Variation	*SS*	*df*	*MS*	*F*	*P-value*	*F crit*
Rows	6049.10	19	318.37	1.11	0.41	2.17
Columns	102.40	1	102.40	0.36	0.56	4.38
Error	5451.60	19	286.93			
Total	11603.10	39				

Figure A.25 The Correlation Between Consumption of Ice Cream and Crime Rate Is .22

	A	B	C	D	E	F	G
1	Community	Crime Rate	Ice Cream Consumption			*Crime Rate*	*Ice Cream Consumption*
2	A	8	36		Crime Rate	1	
3	B	4	22		Ice Cream Consumption	0.22	1
4	C	18	26				
5	D	15	44				
6	E	12	26				
7	F	17	21				
8	G	15	43				
9	H	33	43				
10	I	26	23				
11	J	12	24				
12	K	24	22				
13	L	56	49				
14	M	55	28				
15	N	43	19				
16	O	18	27				

Figure A.26 The Correlation Between Speed and Accuracy Is −.76

	A	B	C	D	E	F
1	Speed	Accuracy			*Speed*	*Accuracy*
2	7	2		Speed	1	
3	5	3		Accuracy	-0.76358	1
4	6	2				
5	7	3				
6	1	7				
7	1	7				
8	3	9				
9	2	2				
10	3	4				
11	4	7				
12	2	7				
13	8	1				
14	7	2				
15	1	8				
16	2	9				

Figure A.27 The Regression Line Is $Y' = -.27X + 82.23$

	A	B	C	D	E	F	G	H	I	J	K	L
1	Wins	Injuries		SUMMARY OUTPUT								
2	77	31										
3	67	34		*Regression Statistics*								
4	87	51		Multiple R	0.21							
5	98	54		R Square	0.04							
6	46	44		Adjusted R Square	-0.08							
7	76	35		Standard Error	20.11							
8	56	65		Observations	10							
9	78	71										
10	74	33		ANOVA								
11	33	64			*df*	*SS*	*MS*	*F*	*Significance F*			
12				Regression	1	147.18	147.18	0.36	0.56			
13				Residual	8	3234.42	404.30					
14				Total	9	3381.60						
15												
16					*Coefficients*	*Standard Error*	*t Stat*	*P-value*	*Lower 95%*	*Upper 95%*	*Lower 95.0%*	*Upper 95.0%*
17				Intercept	82.23	22.51	3.65	0.01	30.31	134.15	30.31	134.15
18				Injuries	-0.27	0.45	-0.60	0.56	-1.30	0.76	-1.30	0.76

Figure A.28 The Regression Line Is $Y' = 4.28X + 121.59$

	A	B	C	D	E	F	G	H	I	J	K	L
1	Sales	Experience		SUMMARY OUTPUT								
2	120	12										
3	143	14		*Regression Statistics*								
4	100	8		Multiple R	0.66							
5	214	21		R Square	0.44							
6	165	15		Adjusted R Square	0.37							
7	242	16		Standard Error	46.34							
8	210	6		Observations	10							
9	222	12										
10	276	33		ANOVA								
11	243	31			*df*	*SS*	*MS*	*F*	*Significance F*			
12				Regression	1	13440.02	13440.02	6.26	0.04			
13				Residual	8	17180.48	2147.56					
14				Total	9	30620.50						
15												
16					*Coefficients*	*Standard Error*	*t Stat*	*P-value*	*Lower 95%*	*Upper 95%*	*Lower 95.0%*	*Upper 95.0%*
17				Intercept	121.59	32.26	3.77	0.01	47.19	195.99	47.19	195.99
18				Experience	4.28	1.71	2.50	0.04	0.33	8.23	0.33	8.23

Figure A.29 A Histogram of Meal Choice by 100 Diners

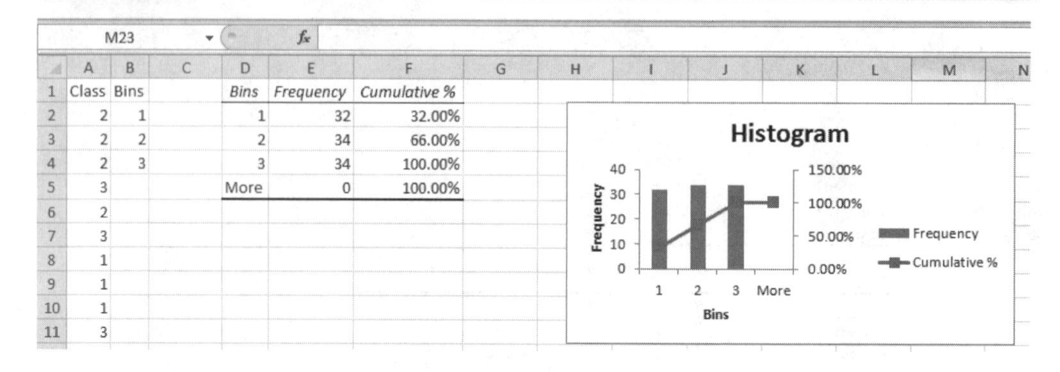

M23

	A	B	C	D	E	F
1	Class	Bins		*Bins*	*Frequency*	*Cumulative %*
2	2	1		1	32	32.00%
3	2	2		2	34	66.00%
4	2	3		3	34	100.00%
5	3			More	0	100.00%
6	2					
7	3					
8	1					
9	1					
10	1					
11	3					

Figure A.30 A Histogram of Activity Choice by 200 Community Members

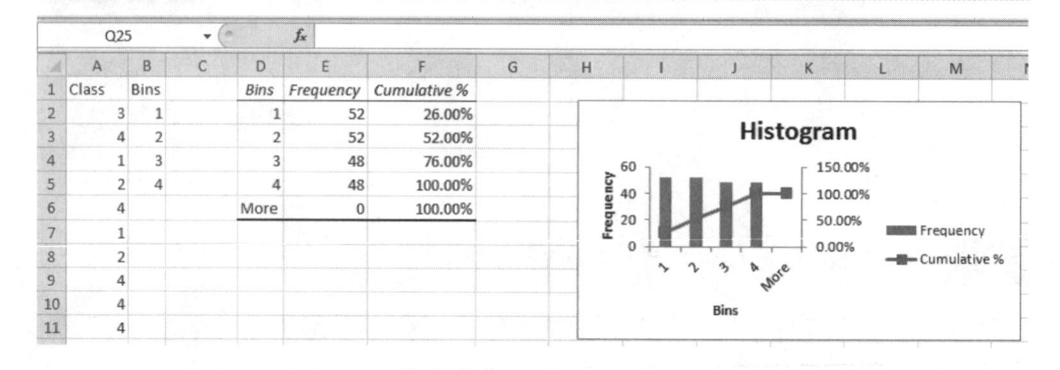

Q25

	A	B	C	D	E	F
1	Class	Bins		*Bins*	*Frequency*	*Cumulative %*
2	3	1		1	52	26.00%
3	4	2		2	52	52.00%
4	1	3		3	48	76.00%
5	2	4		4	48	100.00%
6	4			More	0	100.00%
7	1					
8	2					
9	4					
10	4					
11	4					

Appendix B

Using the Macintosh Excel Formula Builder

For Macintosh users, Excel offers a simple and direct way of using formulas called the Formula Builder. It allows the user to easily enter cell references and return values and can be used to create a formula or enter a specific function.

Figure B.1 shows a set of 10 scores, which we'll use to see the Formula Builder in action. Here are the steps to use the AVERAGE function, the value of which will be returned to Cell B11.

Figure B.1 A Set of 10 Scores to Be Averaged

	A	B
1	7	
2	6	
3	8	
4	5	
5	6	
6	7	
7	8	
8	7	
9	6	
10	7	
11	AVERAGE	

1. Click on the cell where you want the function to appear. In this example, it is Cell B11.
2. Select Insert → Function, and the Formula Builder Window will appear, as you see in Figure B.2.

Figure B.2 The Formula Builder Window

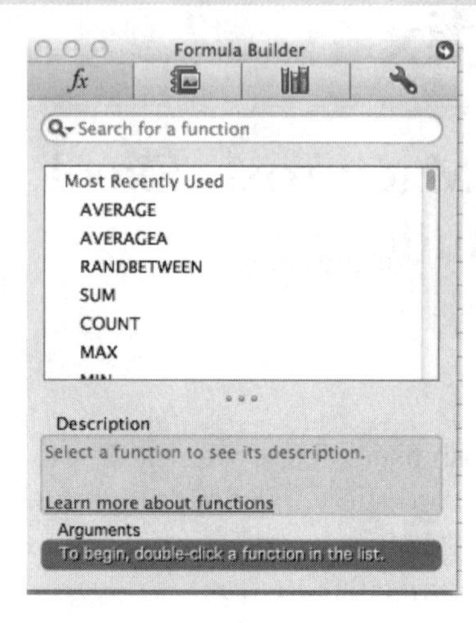

3. You can either search for the AVERAGE function using the search box or scroll down the list to find it. In either case, once you find the function, double-click on it so that the function's arguments appear at the bottom of the Formula Builder window, as you see in Figure B.3.

Figure B.3 The AVERAGE Function and Its Arguments in the Formula Builder

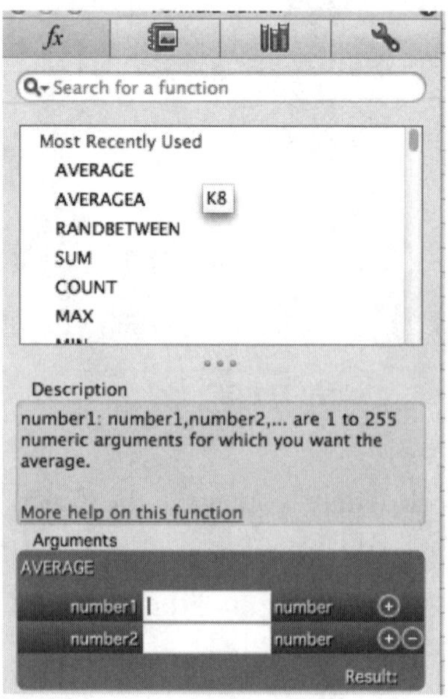

4. Enter the range of values you want to include in the number1 text box. You can enter additional ranges by clicking on the + sign and entering additional cell addresses. In this case, the range for the variable is A1:A10. Press the Enter key. The value of the function appears, as shown in Figure B.4. The average of this set of numbers is 6.7.

Figure B.4 The AVERAGE Function Returning the Average of the Values

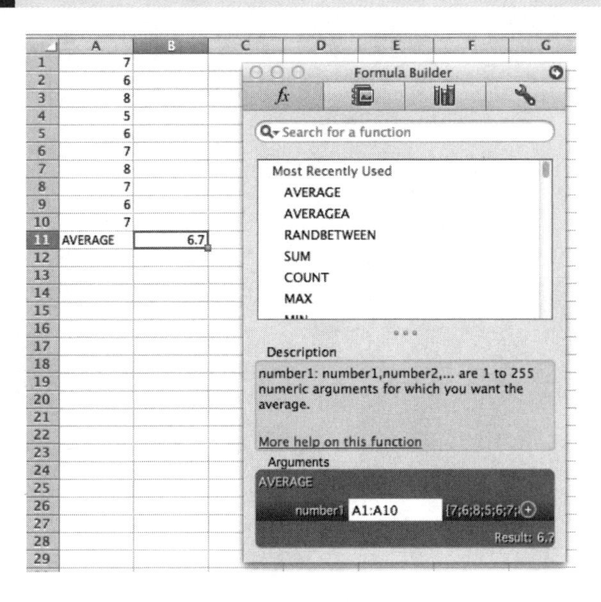

Index